CW00969149

Thinking for a Change

Putting the TOC Thinking Processes to Use

The St. Lucie Press/APICS Series on Constraints Management

Series Advisors

Dr. James F. Cox, III
University of Georgia
Athens, Georgia

Thomas B. McMullen, Jr.
McMullen Associates
Weston, Massachusetts

Titles in the Series

Introduction to the Theory of Constraints (TOC) Management System
by Thomas B. McMullen, Jr.

Securing the Future:
Strategies for Exponential Growth Using the Theory of Constraints
by Gerald I. Kendall

Project Management in the Fast Lane:
Applying the Theory of Constraints
by Robert C. Newbold

The Constraints Management Handbook
by James F. Cox, III and Michael S. Spencer

Thinking for a Change:
Putting the TOC Thinking Processes to Use
by Lisa J. Scheinkopf

Management Dilemmas:
The Theory of Constraints Approach
to Problem Identification Solutions
by Eli Schragenheim

Thinking for a Change

Putting the TOC Thinking Processes to Use

Lisa J. Scheinkopf

The St. Lucie Press/APICS Series on Constraints Management

St. Lucie Press
Boca Raton London New York Washington, D.C.

Published in 1999 by
CRC Press
Taylor & Francis Group
6000 Broken Sound Parkway NW, Suite 300
Boca Raton, FL 33487-2742

Printed in the United States of America on acid-free paper
20 19 18 17 16 15 14 13 12 11

International Standard Book Number-10: 1-57444-101-9 (Hardcover)
International Standard Book Number-13: 978-1-57444-101-7 (Hardcover)

Library of Congress Cataloging-in-Publication Data

Catalog record is available from the Library of Congress

About the Author

Lisa Scheinkopf, CPIM, is a consultant with Chesapeake Consulting, Inc. Headquartered in Severna Park, Maryland, Chesapeake Consulting is well known for its success in helping organizations improve their business results by using TOC and related system improvement technologies. Prior to joining Chesapeake, Lisa was president of the consulting firm, InSync Solutions, where she worked with companies to implement TOC principles and practices. Her clients have included some of the top names in the semiconductor, medical equipment, printed circuit board, and electronics manufacturing industries, the Department of Defense, and several universities. For nearly 20 years, Lisa held marketing, operations management, and materials management positions with such companies as W.L. Gore & Associates and American Socket Screw Manufacturing Company. She did extensive development work and refinement of the TOC strategic thinking processes with Dr. Eliyahu Goldratt when she was associated with the Goldratt Institute. She is considered one of the best in the world concerning the teaching and application of these tools and is a sought-after public speaker on TOC and its applications, systems thinking, and organizational improvement. Lisa is a founding member of the APICS Constraints Management SIG, and served as its chairperson in 1997 and 1998.

Most important, Lisa shares her life with her husband of 18 years, Danny, their two daughters, Jennifer and Rachel, and their not-so-mild-mannered dog, Casey. She can be reached by e-mail at jonalisa@chesapeak.com. Chesapeake Consulting, Inc.'s web site is located at www.chesapeak.com.

ABOUT APICS

APICS, The Educational Society for Resource Management, is an international, not-for-profit organization offering a full range of programs and materials focusing on individual and organizational education, standards of excellence, and integrated resource management topics. These resources, developed under the direction of integrated resource management experts, are available at local, regional, and national levels. Since 1957, hundreds of thousands of professionals have relied on APICS as a source for educational products and services.

- APICS Certification Programs — APICS offers two internationally recognized certification programs, Certified in Production and Inventory Management (CPIM) and Certified in Integrated Resource Management (CIRM), known around the world as standards of professional competence in business and manufacturing.
- APICS Educational Materials Catalog — This catalog contains books, courseware, proceedings, reprints, training materials, and videos developed by industry experts and available to members at a discount.
- APICS — The Performance Advantage — This monthly, four-color magazine addresses the educational and resource management needs of manufacturing professionals.
- APICS Business Outlook Index — Designed to take economic analysis a step beyond current surveys, the index is a monthly manufacturing-based survey report based on confidential production, sales, and inventory data from APICS-related companies.

- Chapters — APICS' more than 270 chapters provide leadership, learning, and networking opportunities at the local level.
- Educational Opportunities — Held around the country, APICS' International Conference and Exhibition, workshops, and symposia offer you numerous opportunities to learn from your peers and management experts.
- Employment Referral Program — A cost-effective way to reach a targeted network of resource management professionals, this program pairs qualified job candidates with interested companies.
- SIGs — These member groups develop specialized educational programs and resources for seven specific industry and interest areas.
- Web Site — The APICS web site at http://www.apics.org enables you to explore the wide range of information available on APICS membership, certification, and educational offerings.
- Member Services — Members enjoy a dedicated inquiry service, insurance, a retirement plan, and more.

For more information on APICS programs, services, or membership, call APICS Customer Service at (800)444-2742 or (703)237-8344 or visit http://www.apics.org on the World Wide Web.

Dedication

We are formed by our past, live in the present, and hope for the future.

This book is dedicated to

The memory of my loving, wise, and courageous father, Sheldon Minow

My husband, Danny, the wonderful man I share my whole being with

And to our daughters, Jennifer and Rachel, the best expressions
of hope and beauty anybody can find.

Acknowledgments

I know every author says that no book is written alone. I am no exception. I am blessed with the support, guidance, ideas, knowledge, love, and friendship of so many. *Thinking for A Change* would not be if it weren't for you. I would especially like to acknowledge:

John Covington for his friendship and gentle disrupting; Dan Hicks for his ability to stay rooted in reality; everybody else at Chesapeake, there is no better group of people in the world to be associated with; Jim Cox and Johnny Blackstone, for teaching the way professors ought to; Dale and Tracey Houle, Christie Latona, and Wendy Donnelly for their encouragement and friendship through all boundary-expanding times; Drew Gierman and the folks at St. Lucie Press for their amazing patience and support; Tom McMullen and everyone involved with the APICS CM-SIG, especially my teammates on the steering committee, all of whom are volunteers, for sticking with the goal of creating, expanding, and disseminating the APICS TOC body of knowledge; and to all my students and clients — I have learned and continue to learn so much from you.

Though we don't know each other personally, I would like to extend my thanks to Margaret Wheatley, Peter Block, Peter Senge, Stephen Covey, and Dee Hock. Your contributions have helped me and many others to understand our organizations as living, learning, growing, ever-evolving systems. And, as such, you have helped put these thinking processes in their place as the learning tools that they are.

Danny, Jenn, and Rachel — any words here would be inadequate to express my appreciation for your support and unconditional love through this project and life.

Finally, my most heartfelt thanks to my friend and mentor, Eli Goldratt. Our journey has led me to better discover and understand myself, grow with confidence, and help others in ways that I never would have predicted.

Men of genius are far more abundant than is supposed. In fact, to appreciate thoroughly the work of what we call genius, is to possess all the genius by which the work was produced.

Edgar Allan Poe, 1844

Contents

PART THREE

The thinking process application tools are quite often combined in order to answer more complex sets of questions. In Part Three, I introduce you to two ways that two or more of the thinking process application tools are combined, providing robust processes for understanding and communicating problems and solutions.

Introduction

When reason and instinct are reconciled, there will be no higher appeal.

Jean-Philippe Rameau, 1734

The Common Denominator

Many manufacturing companies reduce inventory and cycle time, increase customer service, and see their net profits skyrocket.

A small printing company expands into new markets with a product offering that is unlike any combination of products and services it has ever offered before. The firm is rapidly growing market share and profits.

The leadership of one of the largest denominations of organized religion in the world gains an understanding of its key constraint and is in the early stages of implementing a plan that aligns all of the organization with its purpose — to continually improve people's relationships with God and other people.

A man saves his marriage.

A large medical organization within the DoD determines how it will meet the challenges of the end of the century — by providing the readiness it is in business to guarantee, in spite of a continued decrease in personnel and changes in the nature of worldwide military requirements.

A small consulting firm doubles in size every year for five years, while at the same time enhancing its culture, profitability, stability, and reputation in the marketplace it serves.

School children solve their conflicts, study history with vigorous interest, and dramatically improve their grades and test scores.

All of these entities are very different from one another, yet they have some striking commonalities. Each is a system. Each is successful. Each used the Thinking Processes to make the decisions and take the actions that led to the results described above.

In the beginning...

The Thinking Processes were originated by Dr. Eliyahu Goldratt to address the unique and complex issues of firms that were implementing his Theory of Constraints (TOC) in their production environments. Once their manufacturing operations were "fixed," these firms needed more than the techniques described in his landmark text, *The Goal*.

When an organization's physical constraint is a resource in the manufacturing plant, it is in what we consider to be a manageable environment. We're typically dealing with products and machines, and with material that flows within firm, identifiable, presumably controllable boundaries. The data are available, the statistical fluctuations are measurable, the product is touchable, and it's easy to translate the specific steps of the TOC into measures of dollars and time.

When the physical constraint moves outside the physical arena of the plant, it moves into areas that are less controllable, less manageable. Engineering, sales, and marketing issues are unique, a function of the broader mission, culture, strategies, markets, and information systems of an individual organization. Once the material flow is managed correctly, the organization must address deeper issues if it wants to continue to improve.

Just as manufacturing's problems tend to be masked by inventory, the deeper policy and paradigm constraints are hidden by what we tend to call "problems." Just as we look under the piles of inventory to identify the physical constraints that may exist in production, we must look under the "piles of problems" we are confronted with every day to identify the underlying policy and paradigm constraints of the organization — especially when the physical constraint has moved outside the comfort zone of manufacturing. Finding the specific policies and paradigms that inhibit an organization's performance is more difficult than figuring out why a bottleneck is utilized inefficiently.

The simple message of *The Goal* was: clarify the organization's purpose, determine measures that are aligned with that purpose, and improve by managing those few things that limit the organization's higher performance relative to that purpose.

People outside the manufacturing arena also started reading *The Goal*, and its message resonated with many of them, too. While the message was clear, the "how to" wasn't always evident. Even the basic question, "What is the purpose, or goal, of your organization," is often not easy to answer.

Goldratt's Institute was faced with more and more requests from a broader and broader audience. The common theme of the requests was this: *How, specifically, can I apply* The Goal's *ideas to my organization? What, specifically, do I do next? And after that?* The requests came from large organizations such as the Department of Defense and General Motors to organizations as small as Hannah's Donut Shop in Phoenix, Arizona. People from insurance companies, churches, schools, and families were applying lessons from *The Goal* and were seeking the next steps in their journey of ongoing improvement.

Goldratt's challenge was then to avoid becoming the constraint, and he was faced with an interesting dilemma. For people and organizations to truly be successful in their efforts to improve over the short and long term, they must discover and make their own improvements. If this is the case, then the right thing for Goldratt to do would have been to send them on their way, as Jonah had done to Alex in *The Goal*, to figure it all out for themselves. At the same time, in so many cases, it was Goldratt's own ideas that had enabled real turnarounds.

In the vast majority of cases, one of the effects of implementing the Theory of Constraints in a manufacturing environment is the exposure of excess capacity — often to the tune of 30 to 50%! Rather than finding ways to seize the opportunity to grow their markets and their businesses, the short-term and shallow thinking that permeates our culture instead laid off hordes of people. If Goldratt turned people away for the sake of finding their own solutions, and if they didn't then find their solutions, would he be responsible for their companies' inevitable failures? And if he chose to start to solve specific organizational problems, which of those should he choose? Which should he not choose? How would he decide?

The path he chose was neither — or both, depending on how you look at it. Goldratt's solution was to verbalize the processes by which anybody would be able to improve their lot, regardless of the constraints.

Goldratt started to document the thinking processes around 1990. He quickly wrote, *What is This Thing Called The Theory of Constraints,* which conveyed the concept of the thinking processes, and the early versions of the application tools. While it was a great beginning, they were still not easily verbalized and were quite difficult to learn or teach. In early 1991, Goldratt convened several of his partners and associates to tackle this issue. I was honored to be one of the members of this team. Our task was to discover the answer to these questions:

1. If you are "Jonah" (the character from *The Goal*), how do you think?
2. What is the process that you use to really identify constraints, even the invisible constraints, and develop the means to improve any situation, given the existing constraints?
3. How do you teach this process to others?

The team was split into two groups that met over the course of eight months of hard work, sleepless nights, and much discussion, debate, and discovery. In August 1991, the two groups came together in Rotterdam, The Netherlands, to finalize their work. If *What Is This Thing Called The Theory of Constraints* represented the conception of the Thinking Processes, then it is safe to say that in The Netherlands, they were born.

Improvement, or only change?

Every improvement is a change. But every change is not an improvement. The intent of the Thinking Processes was to provide a systematic approach to enable people to create and implement the kinds of change that can also be considered improvement. As such, they were designed to answer these three questions:

What to change?

As we know from the physical world, and the proven effectiveness* of the Theory of Constraints (TOC) as illustrated in Goldratt's *The Goal*, dramatic overall system improvement occurs when we focus our attention on the activities of the system's physical constraint. Does the same hold true in the non-physical world of policies and paradigms?

Think of it this way. How many stupid policies exist in your company? If I asked you to list them, I imagine that your list would be quite long. How do you identify which policies to change? Do you select the easiest policies to target? The policies that are the subject of the majority of complaints? Those that will cost the least amount of money to change? Are you inclined to wipe the slate clean and reengineer the whole shebang?

* Some of the many sources of documented case studies are the *Proceedings from APICS CM-SIG Symposia,* 1995, 1996, 1997, and 1998.

Goldratt was unable to locate any process in existence that looked for the non-physical, invisible weak links — the policy and paradigm constraints. These are the rules and beliefs that, if you change just one or two, you would move your system to a much higher level of performance.

There are, of course, an abundance of techniques that identify root causes, analyze cause–effect relationships, and apply weighting factors to causes in order to prioritize them for improvement projects. These techniques utilize either "hard" data and logic **or** "soft" feelings and intuition. We need processes that utilize the strengths of **both** — processes that enable us to verbalize our intuition and emotions, and as we do so, rigorously test our assumptions surrounding what we've verbalized. With such tools, we can better understand the patterns of our current reality, and select the leverage points that will enable us to inject changes in our systems that will lead those systems to continual improvement relative to their purposes.

To what to change?

Assume that you have found a key policy that is a real constraint to making your organization better than it is today. Now what? Now it's time to figure out what to replace it with. It's also time to consider what you want the new and improved system to look, feel, and act like once its replacement has been implemented. If your current reality is a set of patterns, what would you like the new patterns to be?

Have you ever seen or experienced a solution that didn't work? How about a solution that was just looking for a problem? Or a solution that fixed the problem it was intended to fix, but also caused others? Goldratt sought a process that would enable us to show clearly how a selected solution to a core problem would, in fact, lead to the elimination of that core problem. In addition, he sought a process that would be robust enough to enable us to identify potential, unintended consequences of a solution, and proactively block them. The Thinking Processes were developed in order to have a systematic way to paint a picture of a desired future, in a way that we can determine what we want, why we want it, *and* avoid unintended, undesired consequences that might result.

How to cause the change?

Answering the *What to change* question involves understanding the **current** state of the system. Answering the *To what to change* question involves deciding what we intend the system to be in the **future**. Now it is time to determine how to go about closing the gap. One of the strengths of America as a society is our focus on **action**. We are a nation of doers. My colleagues at Chesapeake have a nickname for this — we are a nation of *gunslingers*. One of our weaknesses, though, is that we are not in the habit of **purposeful action**. All too often, we don't consider what we're trying to accomplish with our actions. We typically don't put much time into considering why we think a specific action will lead to a specific result. We don't ask ourselves often enough if we even need to take an action in order to achieve a specific objective. The result is that we spend a lot of our time doing; we get frustrated because we're working so hard, *doing* so much, and improvement just isn't happening commensurate with the rate of our busy-ness. The Thinking Processes were developed to provide a systematic way to determine what obstacles lie between our current and desired realities, to explain why those obstacles exist, to define the steps that will overcome the obstacles, to define the order in which those steps should be taken, and to recognize when the plan should be altered.

Two Thinking Processes and Five Application Tools

There are really only two TOC Thinking Processes: Sufficient Cause and Necessary Condition. As of this writing, there are five applications of these two thinking processes:

The Sufficient Cause Application Tools	*The Necessary Condition Application Tools*
• Current Reality Tree	• Evaporating Cloud
• Future Reality Tree	• Prerequisite Tree
• Transition Tree	

If you read nothing else, read and do the exercises in the chapters on Sufficient Cause, the Categories of Legitimate Reservation, and Necessary Conditions. You will then be in a position to quickly learn, use, adapt, and modify the five tools — maybe even develop a few yourself!

The Five Thinking Process application tools provide for a systematic approach to answering the three questions for system improvement. The *Current Reality Tree* answers the "What to Change" question. The *Future Reality Tree* answers "To What to Change." The *Evaporating Cloud* is used as a paradigm shifting transition tool between the Current and Future Reality Trees. The *Transition Tree* answers "How to Cause the Change," with the *Prerequisite Tree* serving as the bridge between the Future Reality and Transition Trees.

As with many inventions, uses for the Thinking Processes have evolved into much more than their original intent. These additional uses include

- Communicating
- Decision making
- Resolving conflicts
- Learning
- Developing and implementing policies
- Planning
- Leading and following

As with many things in life, discoveries get made. Original intentions aren't always the lasting intentions of many inventions. This, I believe, may very well be the case with the TOC Thinking Processes. As I have used, taught, played with, challenged, and experienced these tools over the years, I have begun to conclude that their original intent may not be their legacy. These tools provide us with an aid to better and clearer communication with each other as well as better communication between our ears (*thinking*). In an age where agility has quickly become the key to competitiveness, the ability to learn — to generate knowledge — is a true competitive edge. The Thinking Processes provide a systematic approach to increasing our capacity to learn faster and deeper than ever before. The Thinking Processes will prove to be fundamental learning tools in the Age of Agility.

> Learning without thought is labor lost; thought without learning is perilous.
>
> Confucius, 6th Century B.C.

The five application tools may not look the same years from now as they do today. They certainly don't look exactly as they did in 1991. The group that Eli Goldratt brought together in 1991 is today largely dispersed around the world. Many of us have refined and expanded the tools, as have many of our students. I do believe that the principles and the basic thinking patterns on which the five application tools and their variations are based are timeless. Thus, much of this book will be devoted to the principles and basic thinking patterns. Once you've mastered these basics, two things will happen. The first is that the five application tools and variations you might encounter will be very easy to learn and master. The second is that you won't really *need* to learn or master the five application tools, because they will simply be a natural application of the principles and basic thinking patterns. You will likely develop variations of your own.

My cognizant involvement with and use of the Theory of Constraints began in 1987, when I read *The Goal*. The journey has been and continues to be one of nonstop learning and growth. Welcome to the journey.

PART ONE

Chapter 1

The Theory of Constraints

> The whole history of science has been the gradual realization that events do not happen in an arbitrary manner, but that they reflect a certain underlying order, which may or may not be divinely inspired.*
>
> Stephen W. Hawking, 1988

All of TOC, including the Thinking Processes, is based on some fundamental assumptions. This introduction to TOC will provide you with a foundational paradigm that, when adopted, will enable your use of the Thinking Processes to be much more effective.

Resizing the Box

Imagine that I am a new employee in your organization, and it's your job to take me on a tour in order to familiarize me with the company's operations. What would you show me, and what would I see? Perhaps a scenario like this:

> First, we enter The Lobby and meet The Receptionist. Next, we walk through the Sales Department, followed by Customer Service, Accounting, Engineering, and Human Resources. Then, you lead me through Purchasing and Production Control, fol-

* Hawking, Stephen W., *A Brief History of Time,* Bantam Books, 1988.

> lowed by Safety, Quality, Legal, and don't forget the Executive Offices. You save the best for last, and we go on a lengthy tour of Manufacturing. You point out the Press Area, the Machine Shop, the Lathes, the Robots, the Plating Line and Assembly Area, the Rework Area, and the Shipping and Receiving Docks.

Do you notice the *functional* orientation of the tour? I've been led on well over a thousand imaginary and real tours. They are almost always functionally oriented.

Imagine now that we have an opportunity to converse with the people who work in each of these functional areas as we visit them. Let's ask them about the problems the organization is facing. Let's ask them about the "constraints." All will talk about the difficulties they face in their own function and will extrapolate the problems of the company from that functional perspective. For instance, we might hear:

- **Receptionist:** People don't answer their phones or return their calls in a timely manner.
- **Sales:** Our products are priced too high, and our lead times are too long!
- **Customer service:** This company can't get an order out on time without a lot of interference on my part. I'm not customer service, I'm chief expediter!
- **Human resources:** Not enough training!
- **Purchasing:** I never get enough lead time. Engineering is always changing the design, and manufacturing is always changing its schedules.
- **Manufacturing:** We are asked to do the impossible, and when we do perform, it's still not good enough! Never enough time, and never enough resources.
- And so on.

What's wrong with this picture? Nothing and everything. Nothing, in that I'm certain that these good people are truly experiencing what they say they're experiencing. Everything, in that it's difficult to see the forest when you're stuck out on a limb of one of its trees.

My dear friend and colleague John Covington was once asked how he approached complex problems. His reply was, "*Make the box bigger*!" This is exactly what the TOC paradigm asks us to do. There is a time for looking at the system from the functional perspective, and there is a time for looking at a bigger box — from the "whole system" perspective. When we want to understand what is constraining an organization from achieving

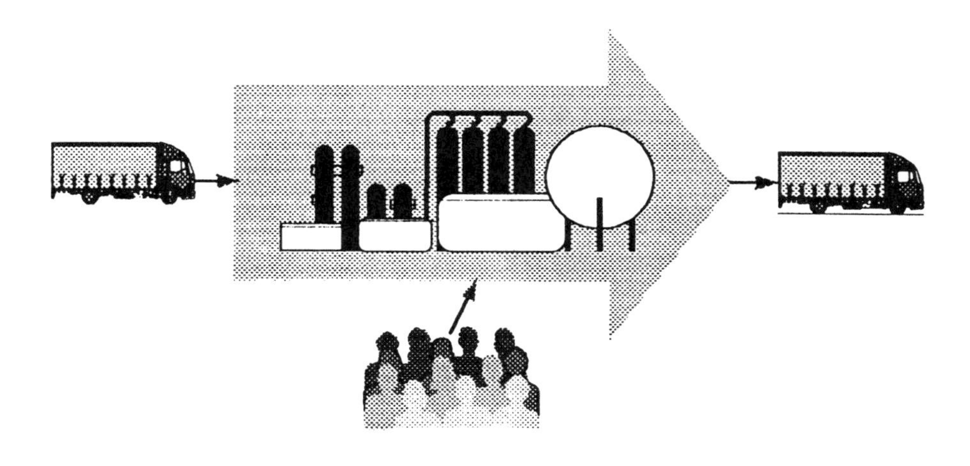

Figure 1.1 The 40,000' Perspective, a.k.a. The Value Chain Box (From *Chesapeake Consulting's Learning and Implementing Change Workshop,* Chesapeake Consulting, Inc., Severna Park, MD. With permission.)

its purpose, we should enlarge our perspective of the box from the Function Box to the Value Chain Box.

Let's take a look at the Value Chain Box. Pretend we have removed the roof from your organization and, for six months, we hover over it at an altitude of 40,000 feet. As we observe, our perspective of the organization is forced to change. We are viewing a pattern. The pattern is *flow*. You may even describe this flow as *process flow*. Whether your organization produces a single product or thousands, the flow looks the same over space and time, as shown in Figure 1.1. The inside of the "box" represents your organization. The inputs to your organization's process are the raw materials, or whatever your organization acquires from outside itself. Your organization takes these inputs and transforms them into the products or services that it provides to its customers. These products or services are the outputs of the process. Whatever the output of your organization's process might be, it is the means by which your organization accomplishes its purpose. The *rate* at which that output is generated is the *rate* at which your organization is accomplishing its purpose. Every organization, including yours, wants to improve. They key to improving is that rate of output, in terms of purpose (a.k.a. *the goal*).

Actually, we can use this box to describe any "system" we choose. For instance, look again at Figure 1.1. Now, let's say that the inside of the box represents your department. Your department receives inputs from

Figure 1.2 40,000' Perspective of Value Added (From *Chesapeake Consulting's Learning and Implementing Change Workshop,* Chesapeake Consulting, Inc., Severna Park, MD. With permission.)

something outside of it, and it transforms those inputs into its outputs. We can also say that the box is you, and identify your inputs and outputs. By the same token, try placing your customers and/or your vendors inside the box — your industry, your community, your country.

In his book, *The Goal*, Dr. Goldratt emphasizes that we need to look at what the organization is trying to accomplish, and make sure that we measure this process and all of our activities in a way that connects to that goal. For instance, when the organization's goal is to make more money now as well as in the future, our 40,000' view of the organization can be translated as in Figure 1.2.

We buy the inputs to the process from our vendors. We put this money into the manufacturing system (which we pay for on an ongoing basis) in order to convert that money into *more* money. It is actually converted into more money when a customer buys the transformed inputs from us, at a price that's higher than what we've paid for them. The term "value added" can be defined, then, as the difference between the money our customers pay us for the outputs of our system, and the money we've paid vendors for the inputs. We make a profit when we've generated more "value added" than we need to pay for the ongoing operation of our system. Note that it's quite easy for us to talk about for-profit companies because the measure of value added — money — is tangible in our eyes, and it comes in clear and precise units of measure. Money is an accepted and common means to measure value.

The basic financial components of any organization are defined by TOC as follows*:

- **Throughput**: the rate at which the system generates money through sales. This is the term used to describe "value added."
- **Inventory:** all of the money the system spends on things it intends to turn into throughput. This is the money we're paying our vendors for the inputs into our system.
- **Operating expense:** all of the money the system spends to turn inventory into throughput. Operating Expenses include everything from wages and salaries to interest expense, taxes, rentals, and insurance — the money we spend to *operate the enterprise*. You might think of this as the money the system spends *on itself.*

When looking for its constraints, an organization must ask the question, "What is limiting our ability to increase our rate of goal generation?" When the goal of the organization is to continually increase its profitability, improvement is measured as throughput (the rate of money generation) increases.

Believe it or not, not all organizations are built for the purpose of making money. Government agencies, religious organizations, public educational organizations, and the like are in business for reasons other than making money. Although none of these entities can afford to neglect the necessity of funds to fuel their survival and growth, their success is not (or at least should not be) measured in dollars. A good example of this is APICS, a nonprofit association of 70,000+ members. APICS defines its purpose as the creation, expansion, and dissemination of knowledge that will improve resource management in manufacturing, service, and government. While APICS is concerned with earning enough money to further its mission, it measures its success by the quality, content, and accessibility of its body of knowledge and the results its members' organizations achieve by interacting with APICS learning resources.

A constraint is defined as *anything that limits a system's higher performance relative to its purpose.* When we're viewing an organization from the functional perspective, our list of constraints is usually quite long. When we're viewing the organization from the 40,000' perspective, we begin to consider it as an interdependent group of resources, linked by the processes they perform to turn inventory into throughput. *Just as the strength of a chain is governed by its weakest link, so is the strength of an organization of interdependent resources.*

---

* Goldratt, Eliyahu M. and Cox, Jeff, *The Goal.*, 2nd rev. ed., North River Press, 1992.

Constraint Classifications

There are three major categories of constraints: physical, policy, and paradigm. All three exist in any given system at any given time, and they are related. Paradigm constraints cause policy constraints, and policy constraints result in mismanaged or misplaced physical constraints.

Physical Constraints

Physical* constraints are those resources that are physically limiting the system from increasing throughput.** Locating physical constraints involves asking the question, "What, if we only had more of it, would enable us to generate more throughput?" A physical constraint can be internal or external to the organization.

At the input boundary of the system, external physical constraints would include raw materials. For instance, if you are unable to produce all that your customers are asking of you because you cannot get enough raw materials, the physical constraint of your organization may be located at your vendor.

An external physical constraint might also be at the output boundary of the system: the market. If you have plenty of capacity, access to plenty of materials, but not enough sales to consume them, a physical constraint of your organization is located in your market. Most organizations today have market constraints.

Internal physical constraints occur when the limiting resource is a shortage of capacity or capability inside the boundaries of the organization. Although it is easy for us to relate to machines as constraints, today's internal physical constraints are most often not machines, but the availability of people or specific sets of skills needed by the organization to turn inventory into throughput.

Every organization is a system of interdependent resources that together perform the processes needed to accomplish the organization's purpose. Every organization has at least one, but very few, physical constraints. The key to continuous improvement, then, lies in what the organization is doing with those few constraints. In *The Goal,* Dr. Goldratt articulated a five-step improvement process that focuses on managing the physical constraints. These five focusing steps can now be found in an abundance

---

* Also called *logistical constraints.*

** Although I'm using the financial reference, you can easily exchange the word *throughput* here with the phrase *goal attainment.*

of TOC literature, and they form the process by which many organizations have achieved dramatic improvements to their bottom line.

The Five Focusing Steps

A process for ongoing improvement, based on the reality of physical constraints

Step 1. ***Identify*** *the system's constraint(s).*
What, if only the system had more of it, would enable it to increase its rate of goal attainment (e.g., throughput)? What is the physical entity that is limiting the system's ability to improve relative to its purpose? Where is the weak link in the chain? Please note that the word *identify* has proactive as well as reactive implications. To identify the system's constraint also means to *decide* where to place it. Thus, an important strategic question that your organization should answer is, where *should* the system's constraint be located?

Step 2. *Decide how to* ***exploit*** *the system's constraint(s).*
What do you want the constraint resource(s) to be doing, to ensure that the system achieves as much of its goal (e.g., generates as much throughput) as it possibly can, given the current state of its resources? How do we get the most with what we've got?

Step 3. ***Subordinate*** *everything else to the decisions made in steps one and two.*
Make sure the rest of the organization is aligned with the exploitation decisions. The job of *every* person in the organization is to enable the organization to accomplish all that its physical constraint is capable of. This is often the most difficult step to implement.

Step 4. ***Elevate*** *the system's constraint(s).*
Increase the capacity of the constrained resource, enabling the system to attain even more of its goal (e.g., generate even more throughput) than its current optimal capabilities. Think of the organization as a funnel with a one-inch-diameter neck. In the *exploit* step, you are doing everything you can to make sure as much fluid as possible flows through that one inch. In the *elevate* step, you are enlarging the diameter of the neck itself, thus enabling even more fluid to flow through it faster.

Step 5. *Don't allow* ***inertia*** *to be the system's constraint. When a constraint has been broken, go back to step one.*
This involves aligning policies and paradigms with the exploitation and subordination decisions. If the physical constraint changes without changes to the policies and paradigms that govern its management, the system will not achieve as much of its goal (e.g., generate the throughput) that it is now capable of.

Policy and Paradigm Constraints

Policies are the rules and measures that govern the way organizations go about their business. Policies determine the location of the physical constraints, and the way in which they are or are not managed. Policies define the markets your organization serves; they govern how you purchase products from vendors; and they are the work rules in your factory. **Policy constraints*** are those rules and measures that inhibit the system's ability to continue to improve, such as through the Five Focusing Steps.

Policies (both written and unwritten) are developed and followed because people, through their belief systems, develop and follow them. In spite of the fact that our organizations are riddled with stupid policies, I don't think that any manager ever woke up in the morning and said, "I think I'll design and enforce a stupid policy in my organization today." We institute rules and measures because we *believe* that with them, the people in our organizations will make decisions and take actions that will yield good results for the organization.

Paradigm constraints** are those beliefs or assumptions that cause us to develop, embrace, or follow policy constraints. In the 1980s, the people who populated many California companies believed their companies were "defense contractors." This belief enforced their policies to market and sell only to the United States government and its defense contractors and subcontractors. Clearly, they had the capacity as well as a wealth of capabilities that could have been productive and profitable serving non-defense-related industries. Nevertheless, the physical constraint for these companies was clearly located in the market. The result, as this industry shrank (or do you say "rightsized"), was that many of these companies went out of business. Their paradigm constraints prevented them from seeing this until it was too late to change the policies that would have enabled them to expand their markets and grow.

---

* Also called *managerial constraints.*

** Also called *behavioral constraints.*

Here are two examples of physical, policy, and paradigm constraints in action, from the lens of the five focusing steps.

A High-Tech Tale

In the southwestern United States, there lives a company that manufactures high-technology electronic products for the communications industry. In this industry, speed is the name of the game. Not only must a company offer very short lead times for its customers, it also must launch more and more new products at a faster and faster pace. The manufacturing organization does a very good job of meeting the challenge by blending the logistical methods of TOC with cellular manufacturing. However, while manufacturing continues to tweak its well-oiled system, the constraint of the company resides elsewhere.

Step 1. *Identify the system's constraint(s).* When I asked the questions, "What is it that limits the company's ability to make more money? What don't you have enough of? Is there anyplace in the organization that the work of the organization has to sit and wait?" It didn't matter who I asked — from the senior executives through people on the shop floor. The answer was almost unanimous: *Engineering!* After some further checking, we learned that specifically, the constraint was the capacity of the Software Design Engineers. Finding Software Design Engineering capacity was the key to this company's ability to increase its new product speed-to-market, and also to the company's ability to make improvements to existing products (in terms of manufacturability and marketability). Here was the key to this company making more money now as well as in the future. Exacerbating the issue was the fact that these types of engineers are very hard to come by, at least in this part of the country. Companies steal engineers from each other and offer large rewards for referrals. If you are a software design engineer, it is not difficult for you to go from company to company and, over a year's time, raise your salary and benefits by 25%.

Step 2. *Decide how to exploit the system's constraint(s).* The company obviously wanted the software design engineers to be doing software design engineering. After a little observation, the company learned some astonishing news. Would you believe that the software design engineers spent only about 50 to 60% of their time doing software design engineering? No, they were not lazy, goofing off, or playing hooky. They were working, and they were working very hard. In fact, Engineering was the most highly stressed, overworked area of the company. At this point, we asked, "What do the software design engineers do that only they can do, and what do they do that others can do?" Some of the tasks involved in the software design engineering job function included data entry, making copies, sending faxes, attending lots of long meetings, tracking down files, supplies, paperwork, and more. This work, though necessary, could be off-loaded to other people. It meant shifting some people around, and yes, wrestling with one or two policy and paradigm constraints. Policy: *the software design engineer does all of the tasks involved with the work that is designated "software design engineering work."* Paradigm: *The most efficient way to accomplish a series of tasks is for one (resource) person to do those tasks. Person (or resource) efficiency is the equivalent of system efficiency.*

Step 3. *Subordinate everything else to the above decisions.* Still contending with the policy and paradigm constraints identified above, subordination meant that anyone feeding work to or pulling work from a software design engineer was to give that work the highest priority. Software design engineering work was now not allowed to wait for anything or anybody, with the exception of the software design engineers. This meant that if you were a nonconstraint, and you were working on something not connected to software design engineering, when that type of work came your way, you put down what you were doing and worked on the software design engineering work. Then, you went back to the task you were working on before.

Step 4. *Elevate the system's constraint(s).* The company is using two routes to increase its software design engineering capacity. The first is that they now have cross-functional teams responsible for the development and launch of new products. As a result, they are reducing the necessity for much of the tweaking, because the designs are better at considering manufacturing, materials, and market criteria from the onset of the new product project. New, manufacturable, and marketable products are being launched faster than ever before. The policy constraint that they had to break was: *Each functional group does its part in the process and then passes the work to the next group.* Of course, this policy stems from the same efficiency paradigm that was highlighted in the preceding steps. The company is also attacking an additional set of policy and paradigm constraints. Policy: *Hire only degreed engineers.* Paradigm: *The only way to acquire the skills of a software design engineer is through getting the formal degree.* Given the general shortage of software design engineers in the region, the company is putting in place an apprenticeship program. In this program, an interested nonengineer will be partnered with an engineer. Over the course of a couple of years, the apprentice will be able to acquire the software design engineering skills that the company needs through a combination of mentoring by the engineer and some courses. This will enable the engineer to off-load some of his or her work early on, which increases the capacity to do the more difficult and specialized work. It also helps the company to develop the capacity it needs in spite of the external constraints (availability of degreed engineers). At the same time, the program will help the company's people to grow, which makes a very positive impact on the company's culture and loyalty of its employees. People feel good when they are helping and being helped by their peers.

Step 5. *Don't allow inertia to become the system's constraint. If, in the above steps, a constraint is broken, go back to step one.* The constraint has not yet shifted out of software design engineering. The current challenge this company faces is to determine where, strategically, its constraint should be, and plan accordingly. In other words, part of its strategic planning process should be to simulate steps one, two, and three, and implement a plan based on decisions resulting from those simulations.

The Studious(?) Student

My eldest daughter is one smart young woman. She is also one of the warmest, friendliest people you'll ever meet. During the early years of her career as a primary grade (K–2) student, Jennifer did great in school. Her teachers always commented on what a pleasure Jenn was to have in class. Her report cards had the best grades. Because she always finished her work in class, Jenn rarely had homework. She was tested and subsequently entered the gifted student program. At this point, things changed. She began to have problems in school. One day, we received a call from her teacher, who let us know that Jenn was in danger of failing and being dropped from the program. Why? She was not turning in her homework. Let's translate this into the five focusing steps, and then I'll let you know how it all worked out.

Step 1. *Identify the system's constraint(s).* As we ask the question, what limits this system from higher performance relative to its purpose, I find there are a couple of questions I want answers to first:

- What is the system we're trying to improve?
- What's its purpose?

I will talk more about these prerequisites to embarking on improvement activities later in this chapter. In this example, the system we were trying to improve was the Jennifer-the-student system. The purpose of this system was to do well in school, which would be evidenced by passing with flying colors into the next grade. We determined that the constraint of this system was Jennifer. Specifically, the "part" of Jennifer that she allocated to doing homework on a timely basis.

Step 2. *Decide how to exploit the system's constraint(s).* This one was almost a no-brainer. Of course, we wanted Jennifer to spend the time necessary to get her homework done, done well, and turned in on time. We decided that Jenn would do her homework right after she got home from school and had her snack. Here is where we ran into our first policy constraint. The policy that Jennifer followed at the time was: *After school, I come home, have a snack, and then go outside to play with my friends.* No problem! We'll just change the rule, right? If you're a parent, you know this was not easy to accomplish. We had to deal with the assumptions (paradigm) that enforced the policy. These included: *School is a breeze. When you do well inside the classroom, homework doesn't matter. Studying outside of school and doing homework are less important than playing after school.* We also had to deal with the fact that the Jennifer-the-student system is a part of larger systems like the Jennifer-the-person system, and the Scheinkopf-family system!

Step 3. *Subordinate everything else to the above decisions.* This included making sure that Jenn had all the supplies she needed, that the kitchen table was cleared up so that she had suitable space to work in, and that she didn't dawdle or take phone calls during homework and study time. This also included Danny (her dad) and I being available as a resource, in case she got stuck on a problem and needed some help, and it meant that we didn't interfere with her studies, like asking her to do other chores during that time or turning up the volume on the stereo. Do you notice that we're addressing additional policies and paradigms in this step, too, and that we are now forced to accommodate the larger systems of which Jenn-the-student is a part?

Step 4. *Elevate the system's constraint(s).* If homework time continued to be the constraint to Jenn's success in school, we would have made additional changes, such as allocating even more study time or getting her a tutor. This was not the case, however. The homework-time constraint was no longer the constraint as a result of applying the first three steps.

Step 5. *Don't allow inertia to become the system's constraint. If, in the above steps, a constraint is broken, go back to step one.* Over the years, the constraint of the Jennifer-as-student system has shifted. At times, it's been external — the degree to which the school programs and curriculum offered enough of a challenge to keep her interested. At other times, it has shifted back to Jennifer, in the form of homework-study time. As she grows, though, the old solutions don't work, and our challenge as a family is to continue to find new, relevant, workable solutions (exploit and subordinate decisions).

Two Important Prerequisites

After many years of teaching, coaching, and implementing, we have identified two prerequisites to the Five Focusing Steps — or, for that matter, to **any** improvement effort — that are not readily obvious. Sometimes, they're just intuitive. Sometimes, they're ignored because they're difficult to come to grips with. When they're ignored, you run the risk of suboptimization and/or improving the wrong things. In other words, you run the risk of system *non-improvement.*

1. *Define the* ***system*** *and its* ***purpose***. What are you trying to improve? Where are you drawing the lines of the 40,000' view box? Around a company? A department? A supply chain? A team? Yourself? *From where* does it receive its inputs, and *to where* does it provide its outputs? What is the system trying to accomplish? Sometimes, the answer is obvious. In the example of the high-tech company, we had our arms around the whole company, and the answer was pretty obvious. We were looking at the whole company and its goal to "make more money now and in the future."*

* "To make more money now and in the future" is a fine and necessary goal for an organization. It does not state, nor does it imply, that the goal should come at the expense of the organization's employees, customers, or other stakeholders. The Thinking Processes can be used to ensure that while the organization makes money, its stakeholders benefit as well.

More often than not, the answer is not obvious at all. Consider the case of a multibillion dollar, multisite chemical company. One of our projects was to help it improve one of its distribution systems. Before we began to talk about the constraints of their system, we asked the team to develop a common understanding of the role of the distribution system as it relates to the larger system of which it is a part. They considered the 40,000' view of the corporation as a whole and engaged in a dialogue on what the purpose of the distribution system is in that "bigger box." As a result, the team was able to focus on improving the distribution system, not as an entity in and of itself, but as an enabler of throughput generation for the corporation.

2. *Determine the system's fundamental* ***measurements***. What does improvement mean for this system? What are its global measures of success? Of failure? How does the system know whether or not it's performing well? This adds clarity to the first question. For instance, let's say that we've defined the system to be a company and that the purpose of the company is to make more money now as well as in the future.* The question of fundamental measurements asks, "*So* what do you mean by *make money now as well as in the future*?" The answer is, "the relationships among throughput, inventory, and operating expense — namely, net profit and return on assets." You know the company is doing well if, over time, its profitability and return on assets is continually good and getting better, but what are the fundamental system-measures of the distribution system that I referenced above? How does it know that it's doing well? Sure, we can say that ultimately they are the same net profit and return on assets. But these measures don't tell the distribution system whether or not it's fulfilling *its* role. The team identified some basic measures that looked at its impact on the company's constraint, as well as the financial measures over which the system has direct control.

Margaret Wheatley, in her work linking organizations with living systems and the sciences of chaos theory, quantum physics, and field theory,** points out that there are three interconnected dimensions to organizational

* I realize that this statement of purpose is strictly the shareholder view. We'll address the perspectives of additional stakeholders, and enlarging or deepening the statement of purpose, as we approach the Thinking Processes.

** Wheatley, Margaret J., *Leadership and the New Sciences,* Berrett-Koehler Publishers, 1992, 1994. Wheatley, Margaret J. and Kellner-Rogers, Myron, *A Simpler Way*, Berrett-Koehler Publishers, 1996.

improvement. Recognizing that people are the inhabitants and improvers of our organizations, she suggests that these dimensions are:

1. Clarity of purpose
2. Quality of relationships
3. Flow of information

As people together enhance the clarity of their common purpose, their relationships improve. As their relationships improve, they open more and more channels through which information can flow, which helps them continue to enhance their relationships and get clearer and clearer on purpose.

In my work with nonprofit organizations, I have come to the conclusion that the answers to the two prerequisite questions are extremely unclear, and this is the root of most of the problems these organizations contend with. At nonprofits, there is a tendency to believe that the measures are so intangible and that attainment of purpose is such a subjective call, that such measures are simply not discussed. The focus ends up to be on measuring and managing the things we call "tangible," such as money. Improvement projects are implemented to improve numbers — membership and fund-raising (or taxes, or tuition). All too often, these projects are undertaken at the expense of moving forward relative to their purpose. This results in dissatisfied stakeholders, drops in membership, losses of money, and a renewed focus on managing the numbers.

For those of you who are employed by for-profit organizations, guess what? The same problem exists. Unless you're the top management or your pay is directly tied to the profitability of the company, it's difficult to rally around the "money is THE goal" banner. Most people want to spend their time in meaningful ways. When companies encourage their people to enter into a dialogue aimed at discovering and clarifying their common purpose as co-members of an organization, the process of improving the bottom line becomes much easier and more fun.

I am **not** advocating that you spend an inordinate amount of time and effort doing process flow and other such diagrams to articulate these things ever so precisely, before you ever get started on the task of improving the system. I **am** suggesting that when you begin an improvement effort, that you begin it with a dialogue on these important issues. (Assuming that you want ongoing improvement, I suggest that you encourage the dialogue to be open and ongoing.) What is the system that we are trying to improve, what's the purpose of the system, and what are its global measures? This dialogue will help you to take a focused and whole-system approach to your improvement efforts.

In the process of clarifying system, purpose, and measures, you may find yourself struggling with the answers to your questions. Enter the TOC Thinking Processes.

In deploying the five focusing steps, it is inevitable that policies and paradigms will need to change. Often, identifying the policies and paradigms that need changing is a "no-brainer." At least as often, it isn't. Enter the TOC Thinking Processes.

Chapter 2

First Steps

It's the quality of the thinking that counts.

Eli Schragenheim, 1996

Shifting Exclamation Points! to Question Marks?

The TOC Thinking Process (TP) tools are used to recognize, verbalize, challenge, and/or change assumptions — starting with our own. Effective use of the TP requires a different mindset than that which we are accustomed to using.

If you are a participant in the Western culture, you are rewarded when you are right and punished when you are wrong. Even back in kindergarten, the child who raised his hand with "the answer" was the child who got picked by the teacher. The "wrong answer" received a "sorry, wrong answer," and the teacher went on to the next child who was frantically waving her arm, convinced she had the "right answer." If that child did recite the right answer, she received warm smiles and praise from the teacher and admiration (along with some jealousy) from the rest of the class.

As managers, we are expected to make presentations after we've done all the research and answered all the questions. The proposal that is presented which results in a thumbs up, no questions asked, is the proposal we all want to make. Questions or challenges to our proposals are considered a sign that we "haven't done our homework." Questioning

> We credit scarcely any persons with good sense except those who are of our opinion.
>
> La Rochefoucauld, 1665

indicates a lack of knowledge, which is considered a weakness. Knowledge is power, so he who has (or at least appears to have) the most knowledge is often anointed "most powerful."

In *The Fifth Discipline** Peter Senge describes the need for our organizations to evolve into learning organizations, which he defines as

> "organizations where people continually expand their capacity to create the results they truly desire, where new and expansive patterns of thinking are nurtured, where collective aspiration is set free, and where people are continually learning how to learn together."

In a world that persists in getting smaller and smaller, and where the rate of change continues to accelerate, our organizations must evolve into learning, adaptive organizations if they have any hope of surviving, let alone thriving. Every organization — whether it is the corporation that employs you, the house of worship you pray in, the family you are a member of, or the community you live in.

I recently commented to the president of a small printing company, "Brad, your company is implementing TOC at the pace of *your* learning. You implement only what *you* have learned and internalized to the degree that you feel comfortable teaching it to your people." He concurred and began to take action to improve his own rate of learning in order to help his company accelerate its rate of improvement. Perhaps one day he will decide that if he gives his employees the opportunity to "learn first," his own rate of learning (and thus improvement), and his company's, will accelerate faster than he ever dreamed possible.

Organizations are collections of people. It's people who operate the technology, share information, perform processes, and determine purpose. Thus, in order for any organization to become a learning organization, people must become learners. This means that we must become observers,

* Senge, Peter, *The Fifth Discipline,* Doubleday, 1990.

> The acquisition of knowledge always involves the revelation of ignorance — almost **is** the revelation of ignorance.
>
> Wendell Berry, 1983

---

questioners, listeners, and information sharers. My experience is that the Thinking Processes provide an approach to articulating, examining, and learning from our thoughts, observations, and communication in a way that deepens learning, while at the same time picking up the pace of learning.

The key, however, is to be open to learning. Try to shed your *I'm right!* attitude and put on your learning hat. It's time to challenge your own assumptions, explore possibilities that your assumptions prevent you from seeing, and listen to others challenge you in a very rewarding way.

Listed below are three steps that I recommend before using any of the thinking process application tools. When you follow these three steps first, you will save time, and you will use the appropriate tool for the situation. You will be less likely to throw away new ideas before you give yourself a chance to explore them. As a result, your solutions will be much more robust.

1. **Formulate the question**. In *The Haystack Syndrome,* Dr. Goldratt defines information as "the answer to the question asked."* The thinking processes are systematic approaches to help us seek out answers to our questions. Thus, they are systematic approaches to finding information in situations where the data might be overwhelming or confusing. Before you sit down to use one of the thinking processes, ask yourself what question you are trying to answer. This will help you to focus and to avoid "paralysis by analysis."
2. **Choose the appropriate tool.** Each of the thinking processes is suited to answering different types of questions. Just as a hammer is the right tool for pounding a nail into the wall and a screwdriver is the right tool for turning a screw, once you verbalize your question, you will have the opportunity to select the application tool or tools most suited to guide you to the answer (or answers). In each application tool chapter, I provide guidelines for the types of questions the specific application tool is suited to help you with.

---

* Goldratt, Eliyahu M., *The Haystack Syndrome,* North River Press, 1990.

3. **Put on your learning hat**. If you already *know!* the answer, don't bother spending time with the thinking process tools! Likewise, if you don't want to find an answer, don't bother with them either. If you do want to find an answer to your question, make your efforts worthwhile — *put on your learning hat*! Old habits often die hard. If you have trouble with this, a visual reminder might be helpful. Some people actually take a hat or a picture of a hat, and put it in their workspace to symbolize they are wearing their learning hat.

Chapter 3

Sufficient Cause: Effect–Cause–Effect

> So your company is making thirty-six percent more money from your plant just from installing some robots? Incredible.
>
> Jonah, 1984
> from *The Goal*

Sufficient Cause

Sufficient cause is the thought pattern of effect–cause–effect. When we assume that something, simply because it exists, causes something else to exist, we are using sufficient cause thinking. Another way of saying this is that we are using sufficient cause thinking when we assume that something is the inevitable result of the mere existence of something else. Here is an example of sufficient cause thinking:

> You live in Chicago, and it is a cold, winter day. You have just gone outside to start your car. You turn your key in the ignition, and nothing happens. If you're like me, you turn the key again, a little bit harder. Then again, harder still, and one more time, just in case you didn't turn it hard enough already. Guess what? You have just used what I call "passive" sufficient cause thinking. In a flash of a moment, you hypothesized that the reason

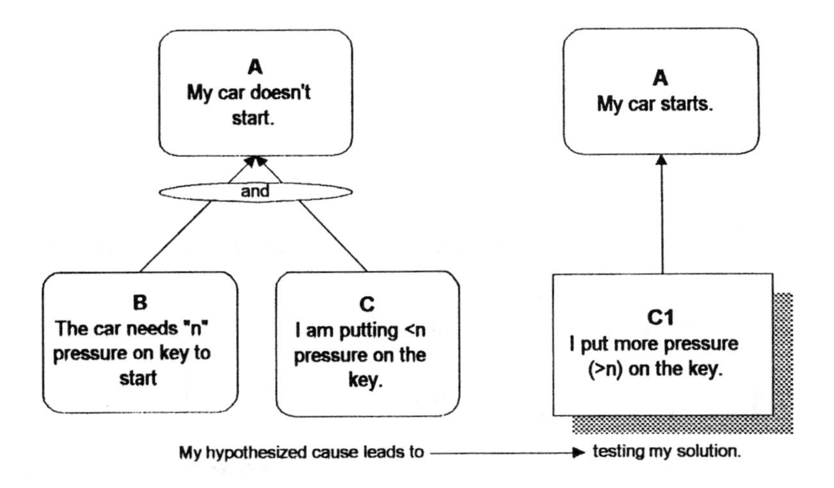

Figure 3.1

your car hasn't started is the lack of pressure on the key. This can be diagrammed as in Figure 3.1. Note the arrow points from your speculated cause (putting less than the required amount of pressure on the key) to its resulting effect (the car doesn't start).

When the only result of your attempted solution is pain in your hand, you realize that the car is not going to start through brute force. You begin to think about what the problem might be, and you move into more "active" sufficient cause thinking. You might guess that the battery is dead, and you begin to check whether or not this hypothesis is correct. You do so by checking for additional effects. Inevitable results of a dead battery would be lights and a radio that don't work, as diagrammed in Figure 3.2.

Notice that you have gone through a pattern of speculating a cause for an effect, and then proceeded to check whether or not you were correct. You checked the validity of your speculated cause by looking for additional, inevitable effects of that speculated cause. *When you are speculating causes for effects, or effects of causes, you are actively using sufficient cause thinking*. The TOC Thinking Processes add a twist, by challenging us to ask *why*. Why do we believe that something causes something else? Why do we believe that an effect is caused by that which we believe causes it?

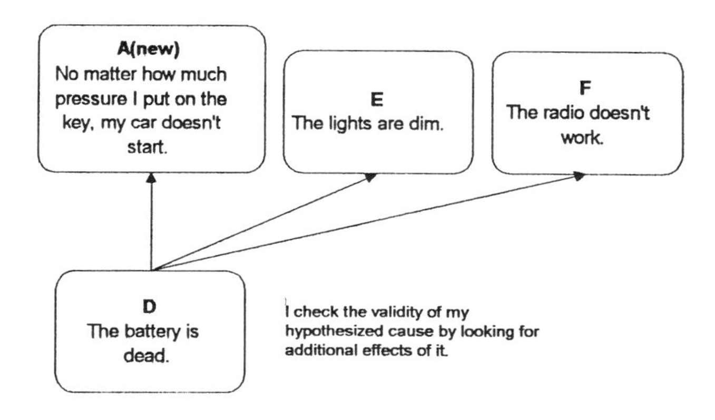

Figure 3.2

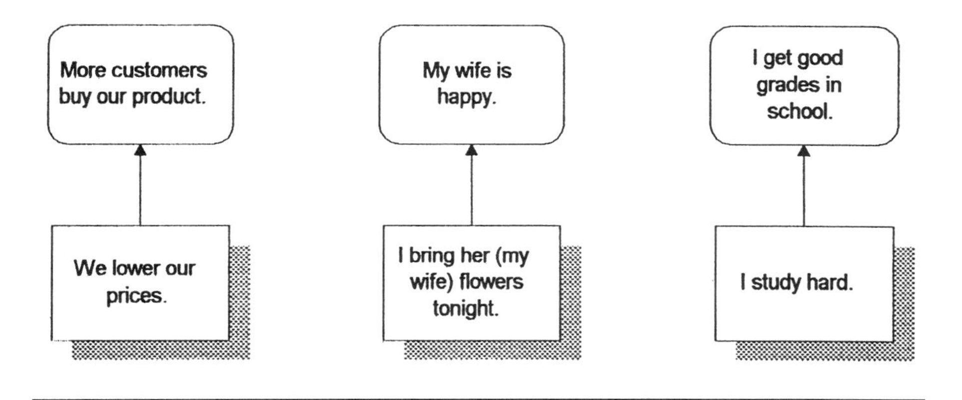

Figure 3.3

We hear and use sufficient cause thinking every day. Some examples are the statements below, which are diagrammed in Figure 3.3:

- *If we lower our prices, then more customers will buy our product.*
- *My wife will be happy if I bring her flowers tonight.*
- *If I study hard, I will get good grades in school.*

Chances are, you're saying to yourself, "So, what's with the diagrams? I learned how to diagram sentences back in grade school. I haven't had a need to do that since then, and I certainly don't see why I should do so now!" The reason for the diagrams is not to separate nouns, verbs, adverbs, and adjectives from each other. The reason for the diagrams is

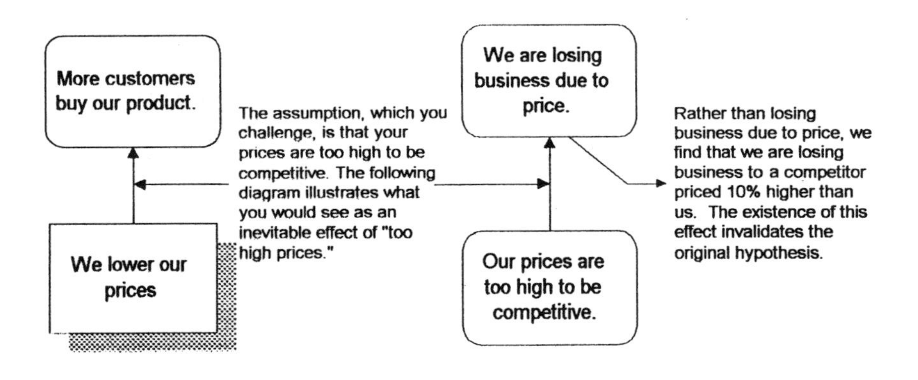

Figure 3.4

that they help us pinpoint the space that might hold hidden, and potentially invalid, assumptions.

"So what?" you might ask. Why care about hidden assumptions? Because hidden assumptions are the source of disagreements and of "best laid plans" gone awry. Identifying the "space" where assumptions hide is the difference between using active and passive sufficient cause thinking.

If you work for a company that needs to increase sales, you have likely heard the statement, *"If we lower our prices, then more customers will buy our product"* (see Figure 3.3). Let's pretend that you are hearing it right now. You might have several concerns. Perhaps you believe that lowering prices will *not* bring in more customers. When pressed to answer why, you point out that your product is already competitively priced, but that you are not competitive when it comes to lead time. In fact, you are losing business to your competitor, whose lead time is two weeks less than yours, and whose price is 10% higher. You have uncovered what you believe to be an erroneous assumption on the part of your colleague — that your product is priced too high to be competitive — and have supported it with evidence of the assumption's invalidity — losing business to a higher-priced competitor. Figure 3.4 illustrates this.

- *How are situations like this typically handled in your environment?*
- *What are the potential ramifications of handling situations in the typical manner?*

Let's continue with this example. Assume now that you do agree with your colleague and believe that if you lower prices, more customers will buy your product. You might still be hesitant to implement a price reduction because you fear that lower prices will result in lower margins, and thus you predict reduced profits for your company. This is illustrated in Figure 3.5.

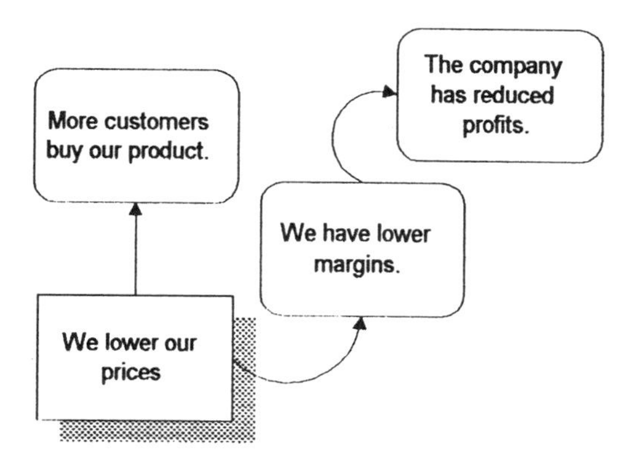

Figure 3.5

What if your job depended on finding out whether lowering prices would likely result in these predicted effects? How would you set about your task? Would you attempt to prove the hypothesis right, or would you attempt to prove the hypothesis wrong? If you are attempting to prove the hypothesis right, you will seek evidence to support it. You might use the diagram to ask the following questions:

- *Why should I believe that more customers will buy our product as a result of lowering our prices? (or, What is it about our reality that leads me to believe that more customers will buy our product as a result of lowering our prices?)*
- *Why should I believe that lower margins will be an inevitable result of lowering our prices? (or, What is it about our reality that leads me to believe that lower margins will be an inevitable result of lowering our prices?)*
- *Why should I believe that lower profits will be an inevitable result of lower margins? (or, What is it about our current reality that leads me to believe that lower profits will be an inevitable result of lower margins?)*

You might choose to take another approach and attempt to prove the hypothesis *wrong*. Your questions would be slightly different:

- *What, if it existed, would force me to not believe that more customers would buy our product as a result of lowering our prices? Does it exist?*
- *What, if it existed, would force me to* not *believe that lower margins will be an inevitable result of lowering our prices? Does it exist?*
- *What, if it existed, would force me to* not *believe that lower profits will be an inevitable result of lower margins? Does it exist?*

The first approach is the approach most of us use today. It is certainly in line with our upbringing in our *"reward the one who's right!"* culture. I suggest that taking the second approach, trying to prove yourself *wrong*, is more effective. Yes, your head might hurt from all the thinking, and your ego might be a bit black and blue as you prove yourself wrong every once in a while. At the same time, you will be exposing those hidden assumptions and learning, while likely saving time and money in the long run. You will implement fewer and fewer solutions that don't solve the problem, and more and more solutions that do!

The Technicalities of Sufficient Cause Diagrams

Sufficient cause diagrams have specific characteristics and terminology. There are seven unique parts to a sufficient cause diagram:

1. Entity (Figure 3.6)
2. Arrow (Figure 3.7)
3. Cause (Figure 3.8)
4. And-Connector (Figure 3.9)
5. Effect (Figure 3.10)
6. Assumption (Figure 3.11)
7. Entry Point (Figure 3.12)

Entity

An entity is a single element of the system. It is expressed as a complete statement. For instance, the phrase, *"the battery"* is not considered to be an entity. However, *"the battery is dead"* is considered to be an entity. An entity can be a *cause* (see Figure 3.8) and/or an *effect* (see Figure 3.10).

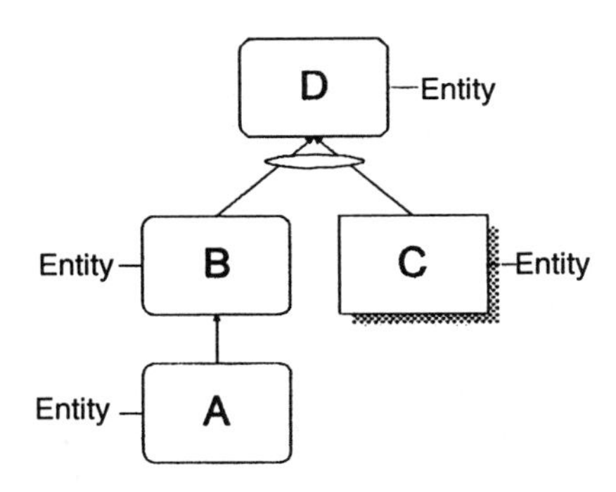

Figure 3.6

Arrow

An arrow is an indicator of a relationship between two entities. The entity at the base of the arrow is the *cause* (see Figure 3.8). The entity at the point or tip of the arrow is the *effect* (see Figure 3.10). An arrow is where *assumptions* (see Figure 3.11) reside.

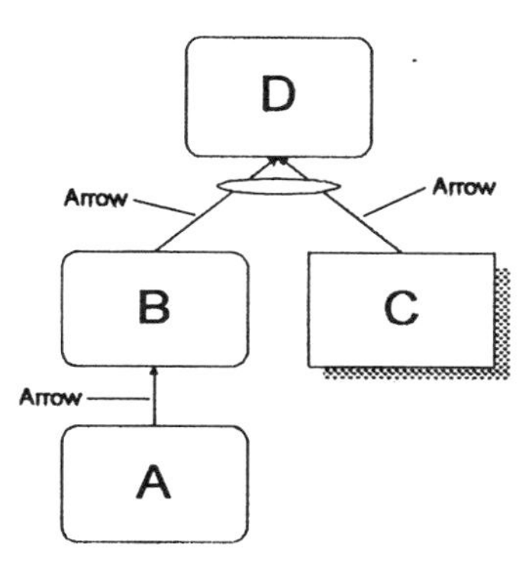

Figure 3.7

Cause

An entity, or group of entities bound by an *and-connector* (see Figure 3.9), that, given its existence, will cause another entity to exist.

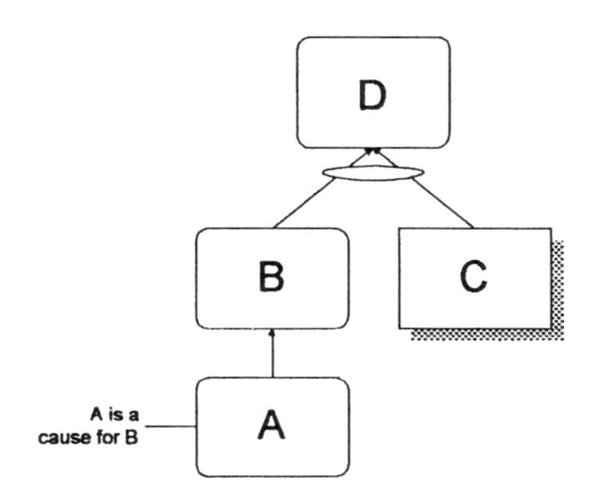

Figure 3.8

And-Connector

The and-connector is an ellipse that groups entities to represent "logical and." Each entity at the base of an arrow that is captured by an and-connector must exist in the system in order for the entity at the point of those arrows to exist as an *effect* (see Figure 3.10) of them.

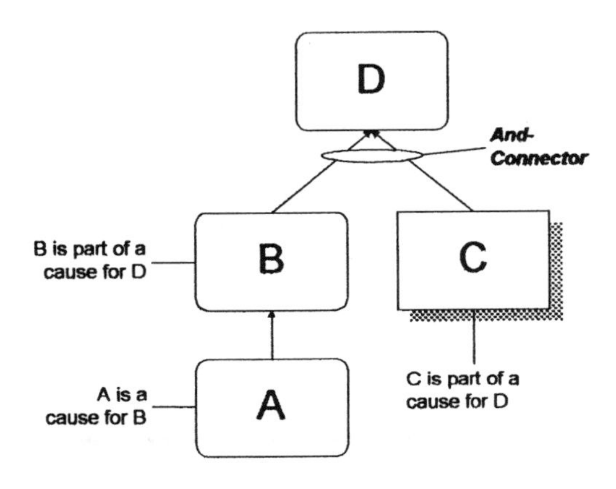

Figure 3.9

Effect

An effect is an entity that exists as an inevitable result of a cause. It is also referred to as a consequence.

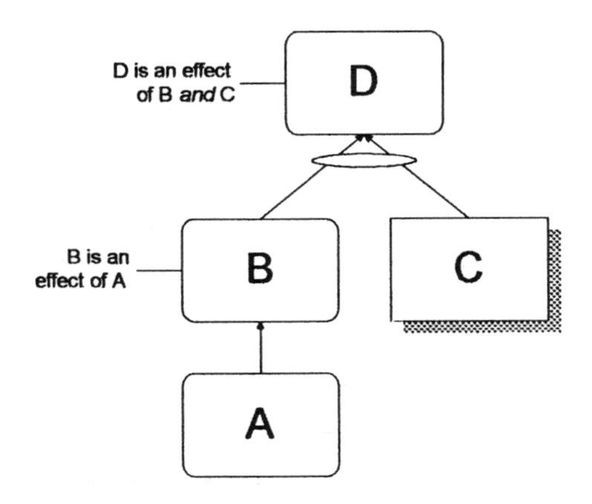

Figure 3.10

Assumption

An assumption is the reason for the existence of the cause–effect relationship. Assumptions lie "underneath" arrows, and are valid or invalid. You can say that the arrow represents the "space" where the assumptions hide. When somebody says, "what's underneath the arrow," they are really asking you to explain the assumptions that connect the cause with the effect.

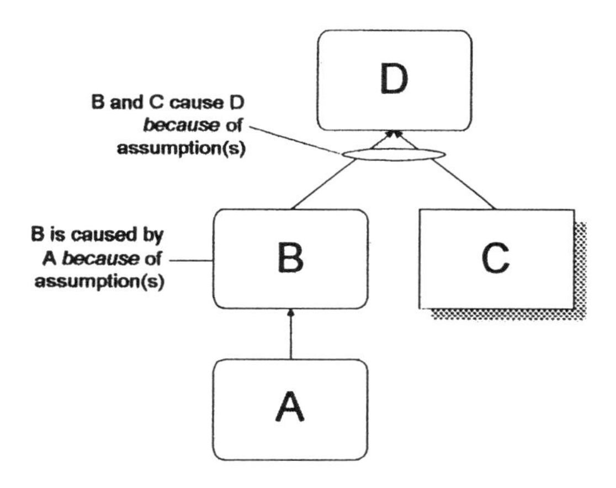

Figure 3.11

Entry Point

An entry point is any entity that does not have an arrow pointing to it. Entry points that are stated in round-cornered boxes (when using a computer program* to diagram) or entry points that are not inside any box (when drawing the diagram by hand) are assumed to exist in the current reality. Square-cornered entry points are entities that do not yet exist. These are called *injections*.

* There are many software products on the market that ease the task of drawing thinking process diagrams. I have created all of the diagrams in this book using Visio flowcharting software by Visio Corporation.

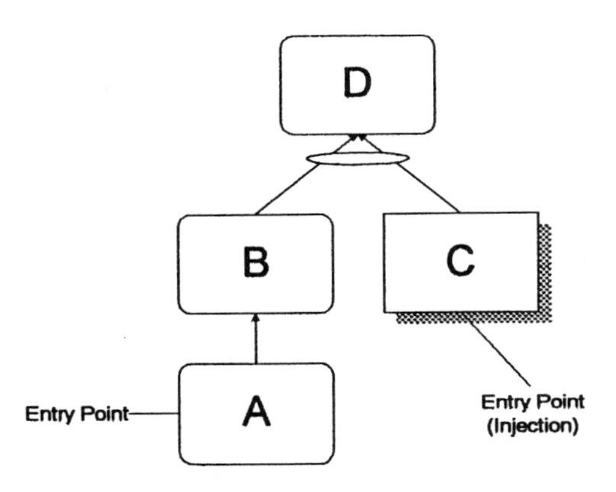

Figure 3.12

Skill Builder: Becoming Aware of Your Use of Sufficient Cause Thinking

1. Choose one instance each day (from your conversations, from your newspaper, TV news program, or other sources) and diagram it as you've seen in this chapter, with causes pointing to effects.
2. Identify the assumptions that formulate the cause–effect relationships that you have diagrammed.
3. Do you agree or disagree with those assumptions?
4. What evidence would indicate that the assumptions are valid?
5. What evidence would invalidate them?
6. Notice how you go about finding the evidence.

Chapter 4

The Categories of Legitimate Reservation

Convictions are more dangerous enemies of truth than lies.

Nietzche, 1878

We use sufficient cause thinking in so much of our thinking and communications day in and day out. Under what circumstances should we stop to check whether or not a claim of cause–effect is valid? Once we make the decision to do so, how do we know when we have checked rigorously enough? Certainly, if we pondered everything, we'd never get anything accomplished!

Several years ago, my husband Danny and I were making the decision to move to another part of the country. The move was going to affect our lives in many ways. Moving our belongings was going to cost several thousand dollars. The area that we were considering had a large population of manufacturing corporations. If I built a large enough local clientele, we reasoned, I would not have to travel as frequently. Our children would be attending new schools. We would have to make new friends and find a new house of worship. The potential impact on us and our family was pretty large, and the risks were significant. This was a decision that would take some thinking, a decision that should not be made "on a whim."

We started our search. We found a general area that we liked, with a synagogue that we felt we would be happy to be members of. There, we asked people about the various neighborhoods, the local schools and school system, etc. We visited several schools and read their reports and ratings. We found a house to rent in a very nice neighborhood. The area was beautiful, and the schools were rated among the top in that part of the state. We had fallen head over heels for the area, so it didn't matter that the rent was going to be three times our mortgage payment in Arizona. We "did the numbers" and predicted that it should be no problem. Business was going to be just fine. We committed to the rental and put our house up for sale. When we moved, the house was still up for sale.

Colleagues, friends, and loved ones expressed their concerns. "Are you sure?" "What if business isn't as great as you think it's going to be?" "This is going to cost so much money!" "Are you sure you can really afford all of this?" "We'll miss you!" I seem to recall that my brother-in-law said something to the effect of, "Are you nuts?" Danny and I thanked them for their concern and assured them that we knew what we were doing. From the moment we began to fall in love with the idea, Danny and I concentrated on all of the reasons to do the move, which helped us dismiss potential reasons not to do the move. We packed our family and our things and off we went.

Our new state was a beautiful place to be, we made many wonderful friends, and, oh, the view from our new home was simply gorgeous! Yet, many of our other expectations never came to fruition. Cost of living was astronomical to say the least, and our optimistic short-term business forecast wasn't exactly accurate. The schools, although "among the best in the area," were significantly worse than those we had left. Our children were receiving little, if any, education. As a result, our youngest was having some behavior problems. In fact, the closer we looked at the published "school report cards," we realized that the school systems in our new state were quite pleased with mediocre academics. Private schools were out of the question. It was hard enough to afford the expensive home, office, and continued upkeep of the home in Arizona, which was now being leased. I won't even begin to elaborate on the "joys" of being a landlord. Less than three years later, we were back in Arizona. A little bit wiser, and carrying our share of financial scars, we were happy to be home.

We, like many of you, operated from the paradigm that tells us to go about checking by looking for all of the evidence that will help us to prove ourselves right. Our motto was, *Justify that decision!* Danny and I thought that we "checked" our hypotheses surrounding our move. We did not check them rigorously, though. *We did not actively try to prove ourselves wrong*. We did not open ourselves to the reservations of our colleagues, friends, and loved ones who, through their doubts, were giving

---

> Man's most valuable trait is a judicious sense of what not to believe.
>
> Euripides, 3rd Century B.C.

---

us hints to the assumptions that we were making that we should have challenged. Instead of looking for the reasons why they might be right, we looked for the reasons why they had to be wrong — so we would continue to be right. If we had allowed ourselves to take another approach, we might have still made the move, but I suspect we would have been better prepared, and we might have taken more time to prepare.

I wonder how much time we spend putting out the fires we started ourselves, because we don't take the time to examine and challenge our assumptions about what we expect to happen and why we had those expectations, especially when those examinations and challenges are handed to us on a silver platter by our peers, team members, subordinates, family, friends, and others with an opinion.

If we are trying to transform our organizations into learning organizations, and if we are trying ourselves to be better and faster learners, shouldn't we be changing these common practices?

Some Guidelines

We don't have to acquire the disease, "paralysis by analysis," nor do we need to bury our heads in the sand. The greater the risk, the greater the need to check, as illustrated in Figure 4.1. I'm talking about the risk of impact on any of the system's stakeholders, not just financial risk. The greater the degree to which the results of your hypothesis might be changing someone's life, the stronger is the need to check.

If the situation calls for checking your claims of cause–effect, use Figure 4.2 as your guideline for how rigorously you should challenge your thinking. When the level of agreement is very low or very high, rigorously seek and challenge the assumptions! This means that if everyone involved is in total agreement right from the start, challenge! You will avoid the dangers of "group think." If there is a good deal of "argument," scrutinize! You will unearth erroneous assumptions and make the appropriate corrections. You will have a much better chance at gaining consensus as well. Now, when I say "everyone," I don't mean everyone you can gather

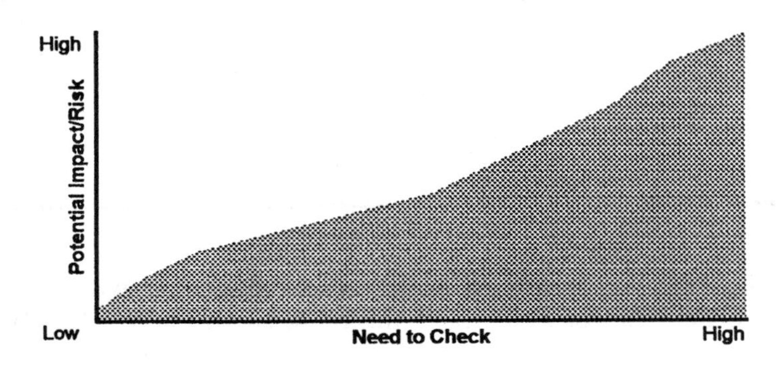

Figure 4.1

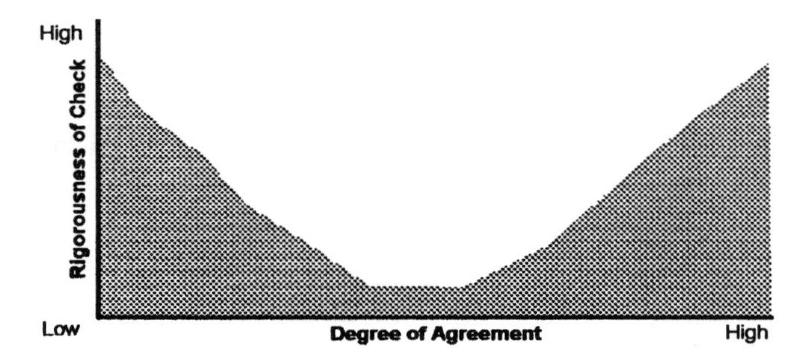

Figure 4.2

around you who already thinks the same way you do. I mean gather a diverse group of stakeholders — individuals who represent different elements of your system, at least some of whom don't think like you. These are the people who have the best chance of helping you uncover those assumptions that hide in the space occupied by the arrows!

Any claim of cause–effect should be able to pass three tests:

1. Verification that what we say exists does exist.
2. Validation of the relationships between causes and their effects.
3. Agreement that what we say reflects what we mean to the people we are attempting to communicate with, including ourselves.

> The fact that an opinion has been widely held is no evidence whatever that it is not utterly absurd; indeed in view of the silliness of the majority of mankind, a widespread belief is more likely to be foolish than sensible.
>
> Bertrand Russell, 1929

Notice the absence of the words "true" or "right." The sciences have taught us that we can't prove anything beyond all doubt — we can test for validity, using our assumptions, and act based on what we've learned to date. We verify, validate, and invalidate. We expose our assumptions and test them against reality as we know it. When assumptions are found to be valid or invalid, we learn quite a bit, and adjust what we do accordingly.

The TOC Thinking Processes provide a methodology for testing our claims with a set of tools called the Categories of Legitimate Reservation (CLR). These tools blend basic scientific method with the powerful intuition we gain through our life experiences to provide an easier, yet systematic approach to uncover, verbalize, challenge, and replace our assumptions while using sufficient cause thinking. The Categories of Legitimate Reservation are illustrated in Figure 4.3.

The current reality, future reality, and transition trees depend on your rigorous use of these categories of legitimate reservation. Master them, and several things will happen:

- The quality of your thinking will improve
- The quality of your listening and communicating will improve
- You will have established the foundation to master the sufficient cause application tools quickly and use them very effectively

Level One Reservations

> All the mind's activity is easy if it is not subjected to reality.
>
> Marcel Proust, 1913

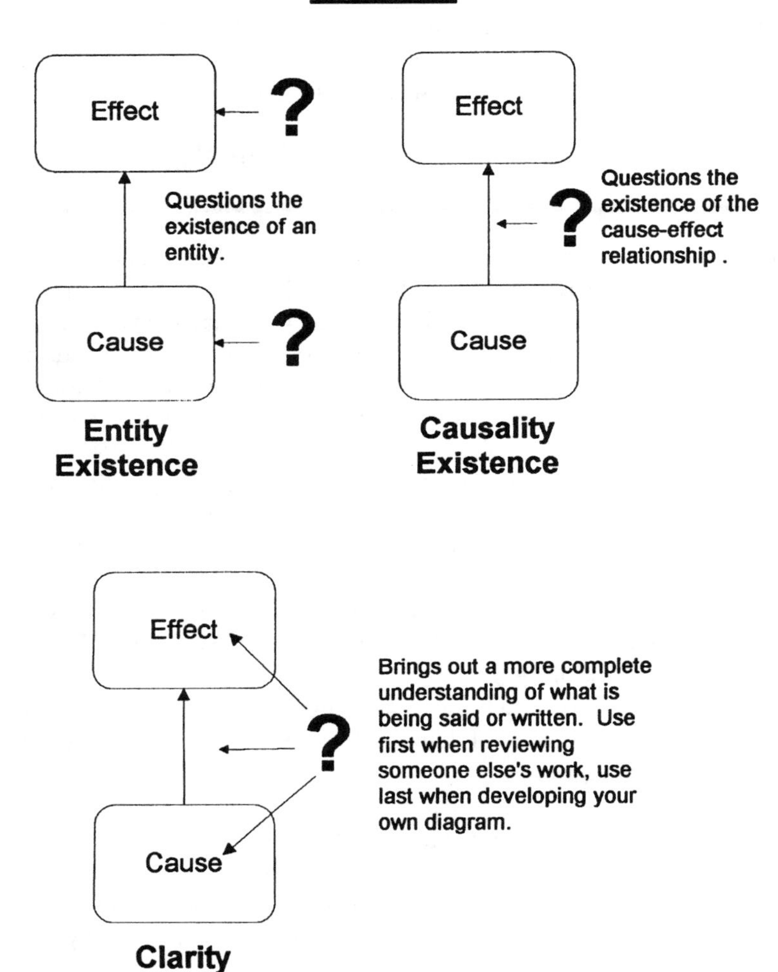
Categories of Legitimate Reservation
Level One
Effect
?
Questions the existence of an entity.
Cause
?
Entity Existence
Effect
Questions the existence of the cause-effect relationship .
?
Cause
Causality Existence
Effect
?
Brings out a more complete understanding of what is being said or written. Use first when reviewing someone else's work, use last when developing your own diagram.
Cause
Clarity

Figure 4.3

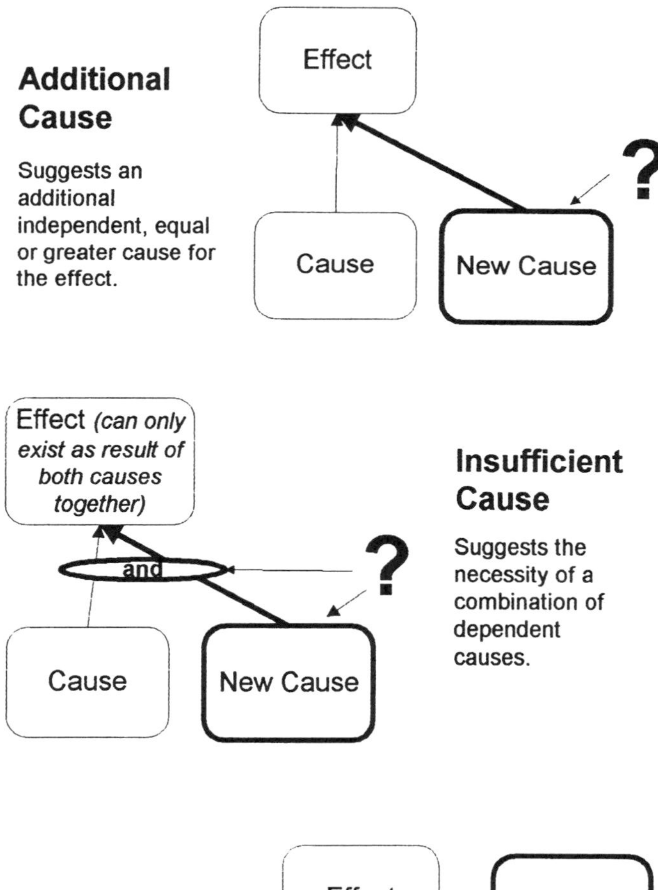

Predicted Effect

Tests the existence of a cause and/or existence of an effect by predicting an additional effect that must also result from the cause.

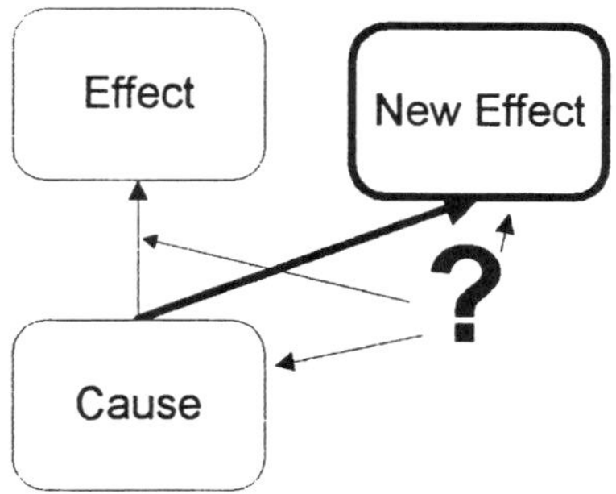

Figure 4.3 (continued)

The first three categories are what I consider to be the basics, because they directly ask the three fundamental questions of sufficient cause.

1. *Entity* Existence asks us to verify that what we say exists does exist.
2. *Causality* Existence asks us to validate relationships between causes and their effects.
3. *Clarity* asks us to make sure that what we say reflects what we mean to the people we are attempting to communicate with, including ourselves.

Entity Existence

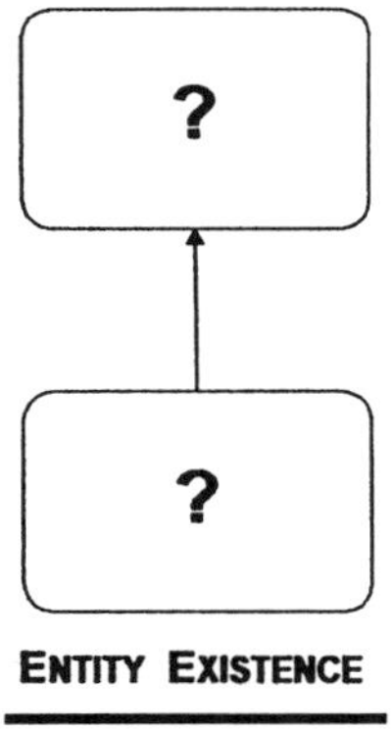

In the previous chapter, I described the various elements of sufficient cause diagrams. An *entity* is a single element of the system, expressed as a complete statement. The *entity existence* reservation is used to verify whether or not an entity does, in fact, exist in the system that is being examined. *Is it really there?*

?

?

ENTITY EXISTENCE

Figure 4.4

Let's take a look at a simple example of entity existence. Perhaps you will relate to this scenario:

Mother to daughter: *Daughter, is your room clean yet?*

Daughter, avoiding eye contact with mother: *Yes, Mom. My room is spotless, and I'm ready to go to the mall.*

Mother is not convinced that Daughter's room is really all that clean. *Are you sure, Daughter, that your room is **spotless**? Shall I go take a look?* In terms of the Categories of Legitimate Reservation, Mother has just expressed an entity existence reservation.

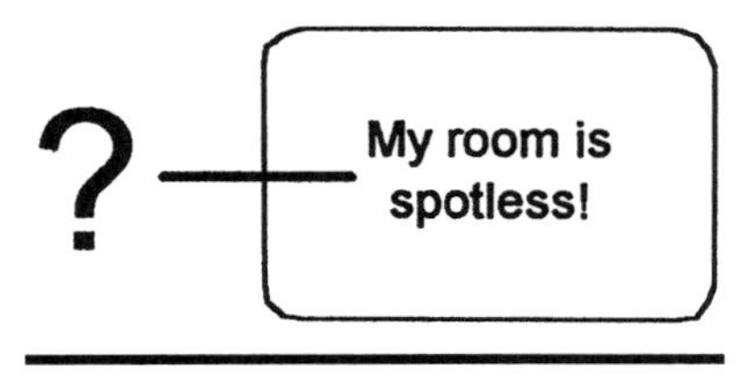

Figure 4.5

Daughter: *Well, er, um, maybe it's not quite spotless, Mom. It is a lot cleaner than it was, though! I need to make my bed and dust my dresser. I'll let you know when I'm finished.*

A couple of months ago, I went on a business trip to Denver. After I arrived at the Denver airport, I was to take a shuttle bus to my hotel. I was tired and hungry, and the wait for the shuttle bus seemed to take forever. I finally arrived at the hotel, where my room was about as far from the lobby as a room could be. I turned up the light switch as I entered the room. No lights came on. I went farther into the room, and turned on the lights over the dresser and by the bed. No problem. I then tried to turn on the lamp next to the table. No light. I looked at the bulb. It had that dark, used-up look to it. I then made sure it was turned tightly in its socket. Although the bulb was tight in the socket, the socket itself was loose and was turned on its side. *Aha!* I thought. Must be a short, or the bulb itself is out. I needed that lamp, because I had work to do that evening. I called the front desk, and they sent a gentleman from maintenance to fix the problem. When he got to my room, I told him, "The lamp doesn't work." Before he attempted to fix it, he did something very important. *He had an entity existence reservation* and checked to make sure that the lamp, in fact, wasn't working. He went over to the wall, found the unplugged cord, plugged it in, and turned on the lamp. Guess what? The lamp worked. (Talk about an embarrassing moment! Is it OK with you if we chalk it up to being tired?)

The lamp doesn't work.

Figure 4.6

The entity existence reservation reminds us that we should check to make sure (verify) a problem exists before we spend time figuring out what's causing it or how to solve it. As illustrated in the car battery example of the previous chapter, it also reminds us that when we have speculated a cause for something, we should make sure (verify) that those (cause) entities also exist as entities in the reality we're trying to understand. (*Sure, a dead battery can be a cause for a car failing to start, but is it so in* ***this*** *case? Use the entity existence reservation, and check first to see if the battery is dead!*)

Warning: Don't make this more difficult than it is. The entity existence reservation is used merely to remind you to ask yourself a simple question: Does this really exist? If the answer is yes, fine. If the answer is no, excellent! You've saved yourself some work in the long run, because you won't be solving nonexistent problems. If the answer is "I don't know," move on to the predicted effect reservation later in this chapter. Its purpose is to help you systematically check those entities in particular.

Causality Existence

Do babies *really* come from storks?
Does smoking *really* cause cancer?
Will customers *really* buy our product if we lower our prices?

The *causality existence* reservation is used when you are questioning the cause–effect relationship itself. In terms of the sufficient cause diagram, the causality existence reservation points directly to the assumptions that occupy the space underneath the arrow, arrows, or and-connector. It does not question the existence of any of the entities, but rather the cause–effect *relationship* that is hypothesized to exist between them. The causality existence reservation says, *OK, I believe the entities exist, but does the (speculated) cause* ***really*** *cause the effect? Does the effect* ***really*** *exist simply because the (speculated) cause does?*

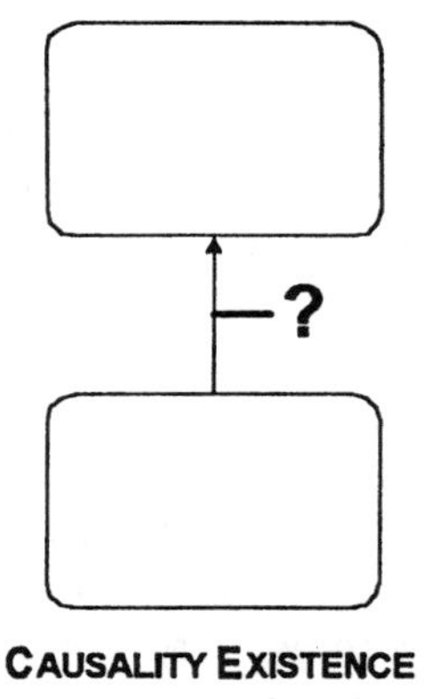

Figure 4.7

I hear thunder.

I see lightning.

Figure 4.8

Let's say that someone showed you Figure 4.8. What is your initial reaction? At first glance, you might say, "Yes, I agree, thunder does follow lightning." Take another look and evoke the causality existence reservation. Assume that both of the entities exist, exactly as they are stated: I hear thunder. I see lightning. Does the fact that one sees lightning really cause that person to hear thunder? Is hearing thunder an *inevitable result* of seeing lightning? One might argue that just because thunder tends to be audible after lightning occurs, doesn't mean that I will hear thunder *because* I see lightning. Or does it? Figure 4.9/11 illustrates a few ways in which this diagram might be changed, *depending on the realities of the situation*, as a result of applying the causality reservation. This simple example helps us uncover two common behaviors that often hinder our practices of thinking and communicating.

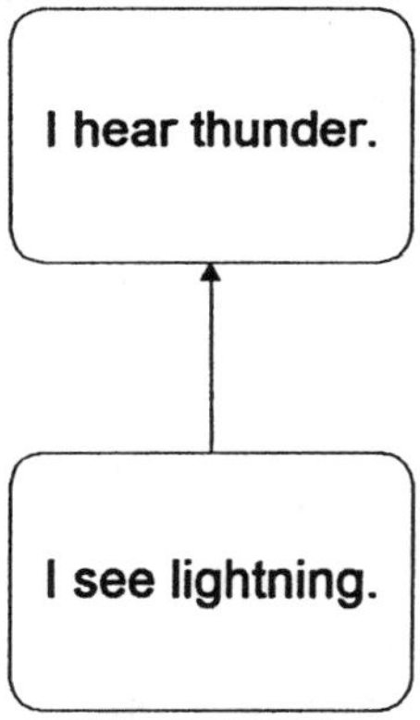

1. We have a tendency to confuse the sequential occurrence of events over time with cause and effect. In a sufficient cause diagram, every arrow denotes cause–effect. This is a significant departure from a process flow diagram, in which the arrows typically signify sequence.
2. We always perceive things first from the perspective of our own assumptions. It is the meaning that we attach to what we hear, see, or read, that gives rise to agreement or disagreement. Whatever your first response to the thunder and lightning example happened to be, that response came from the assumptions that you made about the diagram that you read.

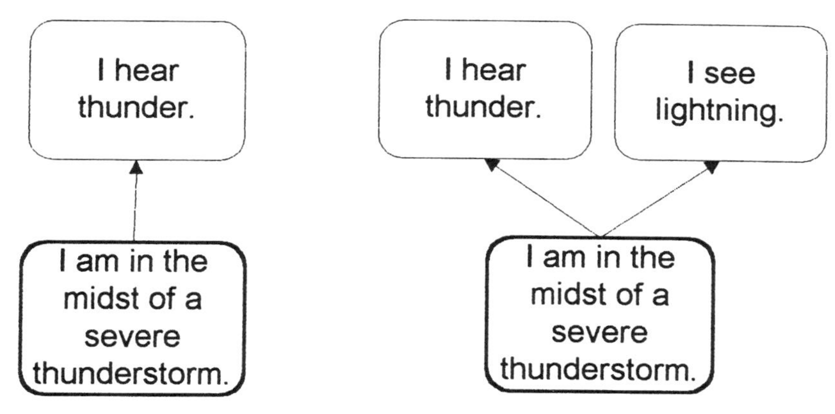
I hear thunder.
I am in the midst of a severe thunderstorm.
In this case, the original entity that was the speculated cause is replaced with a different entity.
I hear thunder.
I see lightning.
I am in the midst of a severe thunderstorm.
In this case, the original entity that was the speculated cause is replaced with a different entity. It also accounts for the entity that was the originally speculated cause, and is diagrammed as such.

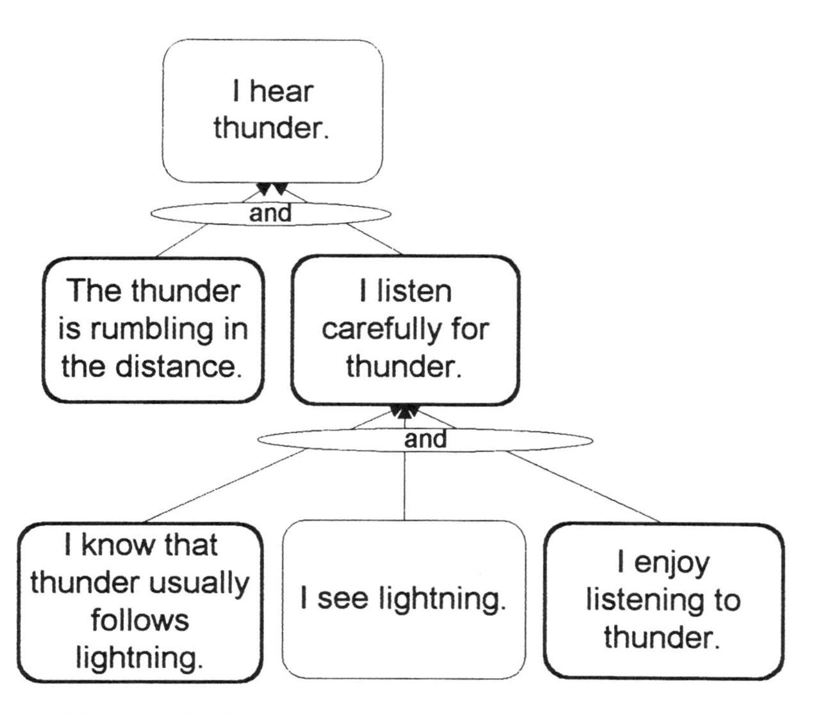
I hear thunder.
and
The thunder is rumbling in the distance.
I listen carefully for thunder.
and
I know that thunder usually follows lightning.
I see lightning.
I enjoy listening to thunder.
In this case, entities have been added to show how seeing lightning is, in fact, part of what causes me to hear thunder.

Figure 4.9/11

The causality reservation gives us an opportunity to test our assumptions against the reality of the validity of a cause–effect relationship — a relationship that we are claiming, as well as one that we are reading or hearing.

Warning: Don't make this more difficult than it is. The causality existence reservation is used to remind you to ask yourself a simple question: *Does this cause–effect relationship really exist?* If the answer is yes, fine. If the answer is no, excellent! You've saved yourself some work in the long run, because you won't be solving the wrong problem. If the answer is "I don't know," move on to the Additional Cause and/or Cause Insufficiency reservations later in this chapter.

Clarity

I continue to be impressed at how often things go wrong simply because of a lack of clarity. For instance, have you ever found yourself in "violent agreement" with a colleague? There are three scenarios in which the clarity reservation is appropriate, and all three of these scenarios occur more often than you think:

1. You are not understood by someone you are trying to communicate with.
2. You do not understand what someone else is trying to communicate to you.
3. You have not articulated your own thinking clearly enough to yourself.

A matter that becomes clear ceases to concern us.

Nietzche, 1886

When an entity is not written as a complete statement, automatically raise a clarity reservation! When you are writing the entity, be sure to write it as a complete statement. It will be there as your own reminder of what you mean. When you leave a diagram and come back to it some time later, you will not have to second guess yourself as to what you meant.

If the entity is one that is being verbalized by someone else, know that the only way you make sense of it is by completing a sentence in your mind. For instance, let's say you are having a conversation with the owner of a small printing company. You can tell just by looking at his face that he's upset about something. You ask what's troubling him. His response, as he shakes his head, is, "Sales." You might assume that his problem is that his company's sales are low or declining. What if I told you that the problem is that his company's sales are too high? You see, they executed a magnificent marketing strategy, and now they are flooded with orders — orders that they don't have the capacity to produce. The executive is upset because he fears that unless his company does something, many of the new customers they attracted will leave because of poor service. The point here is that you really had no idea what he meant by the word "sales" until you heard or made up the remainder of a complete sentence — such as, "We don't have enough capacity to handle all of our new sales."

To "dialogue" is to communicate in a way that expands the pool of meaning or information, with a focus on creating and enhancing shared purpose, mutual respect, and clarity of understanding. One of the principles of dialogue, as taught by The Praxis Group in Provo, Utah, is *Work on me first, us second.* Like Steven Covey's *Seek first to understand before you seek to be understood* *, it reminds me that before I blame or complain about anybody else, and before I assume that anybody else is wrong and I'm right, I had better check inward first. I will admit, I do ignore the principle from time to time, but when I do, I always end up regretting it.

When reading someone else's diagram (or for that matter when listening to someone else speak or reading what they've written), if you are inclined to disagree, assume first that you have a clarity reservation. This means that you assume first that you are not fully understanding what they are saying, before you assume that they're wrong. You will find that as you open yourself up to this practice, you will have fewer arguments and better relations.

Level Two Reservations

The remaining three reservations are used when the questions posed with the entity or causality existence reservations are unanswered. A reservation is unanswered when there is still doubt among any of the parties involved.

4. *Additional cause* asks us to further examine causality existence by looking for additional independent causes for the given effect.
5. *Cause insufficiency* also further examines causality existence by looking for missing dependent elements of the cause.
6. *Predicted effect* is used to examine either causality or entity existence by utilizing the scientific method of effect–cause–effect.

* Covey, Stephen R., *The 7 Habits of Highly Effective People*, Simon and Schuster, 1989.

Additional Cause

Is the new incentive system the *only* reason that morale has improved?

Was the price increase the *only* reason for the decrease in sales last year?

Did Johnny fail his algebra class *only* because he was absent twenty times in the semester?

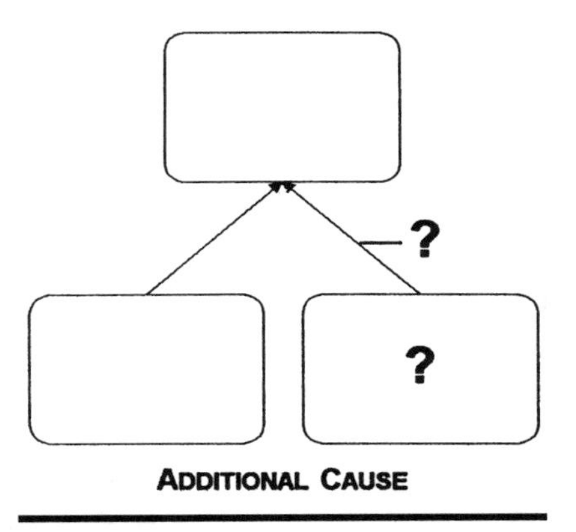

Figure 4.12

The *additional cause* reservation is used when you suspect that the hypothesized cause is not the *only* cause for the resulting effect. A good test for *additional cause* is to ask, *"If we removed the originally speculated cause, would the effect still exist?"* If the answer is yes, there is a good chance that an additional cause exists. Let's use Johnny's case as an example. Johnny failed his algebra course, and his parents want to understand why. A conference took place between Johnny, his teacher, and his parents. When asked why Johnny failed, the teacher told them that Johnny had been absent 20 times in the semester. That alone was sufficient cause for failure, because only five absences were allowed by the school. Together, they develop a communication plan designed to ensure that Johnny attends class. Johnny's dad then asks the teacher, *"Let's assume that our plan works, and Johnny is in class every day next semester. Will he pass, or is there anything else that we should be aware of?"* Johnny's dad expressed an *additional cause* reservation. What else, in addition to the poor attendance, caused Johnny to fail his algebra class?

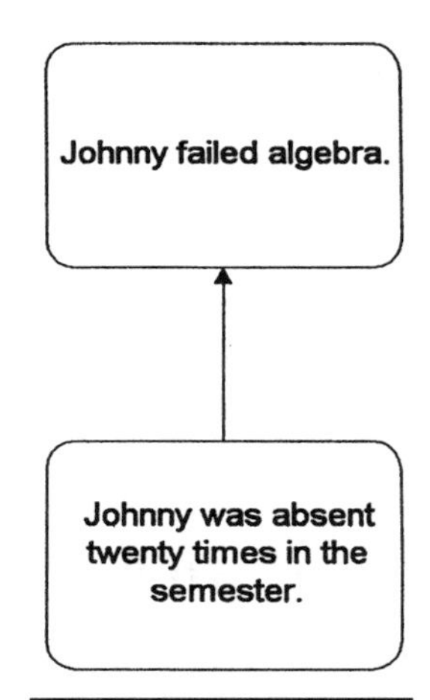

Figure 4.13

The teacher then tells Johnny's parents that Johnny rarely turned in his homework. This, too, was sufficient to earn Johnny his failing grade. The group then expands the communication plan to include making sure Johnny regularly turns in his homework. *Anything else? Are his test scores OK? When he's in class, does he behave?* Johnny's mother raised the additional cause reservation once more, just to be sure. *No,* said Johnny's teacher. *If Johnny improves in these two areas, he should see a passing grade.*

When utilizing the additional cause reservation, we are using our intuition and our experience as we think of other potential causes for the given effect. In her questioning process, Johnny's mother gave examples of other potential causes that she was aware of for failing a course. The only entities that remain on the diagram are the entities that do, in fact, cause the effect in the system or situation that is under examination.

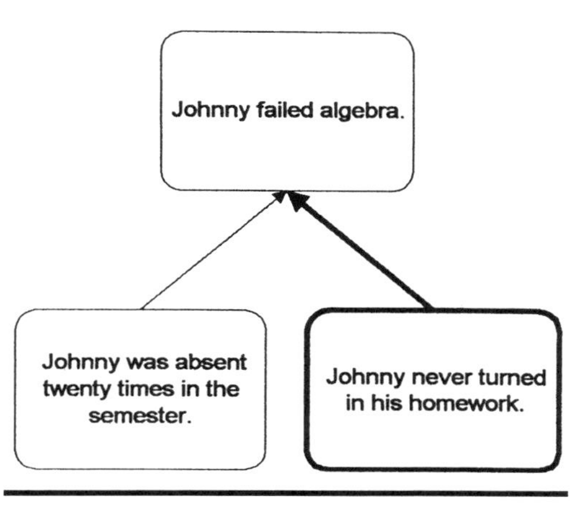

Figure 4.14

The question that you are attempting to answer with the additional cause reservation is, *Is there something else, independent of the cause(s) already speculated or validated, that is causing the effect?* One way to get there, as Johnny's mother did, is to ask two questions:

1. *Is there something else that might cause the effect?*
2. *Does that something else exist in the reality of the situation or system that I'm examining?*

It's easy to overdo this reservation, by adding so many minor causes that the diagram becomes meaningless. For each cause, ask *If we eliminated this entity from existence, to what degree would its effect still exist?* Take a look at Figure 4.15, which is what I refer to as a pincushion. Let's say entity C provides for 50% of Z's existence, entities D and A each provide for 20%, and entities B and E each provide for 5%. If entity Z were something that you wanted to change or improve, working to eliminate B and E will not have much of an impact. In a case such as this, my suggestion is to remove entities B and E from the diagram.

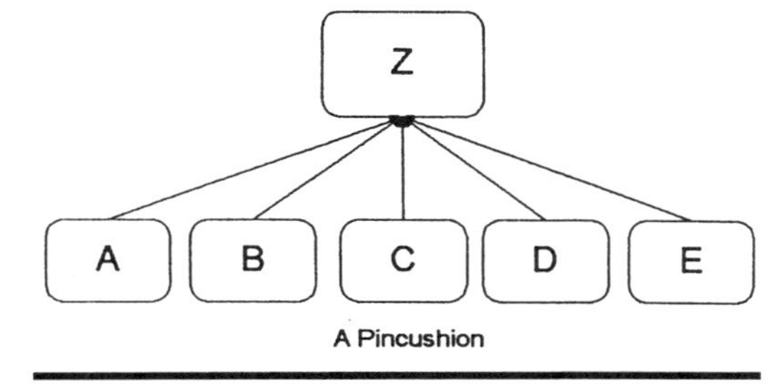

Figure 4.15

Insufficient Cause

The insufficient cause reservation questions causality by identifying one or more entities that must exist along with the speculated cause entity or entities in order for the cause to be valid. The questions that we're asking with this reservation are:

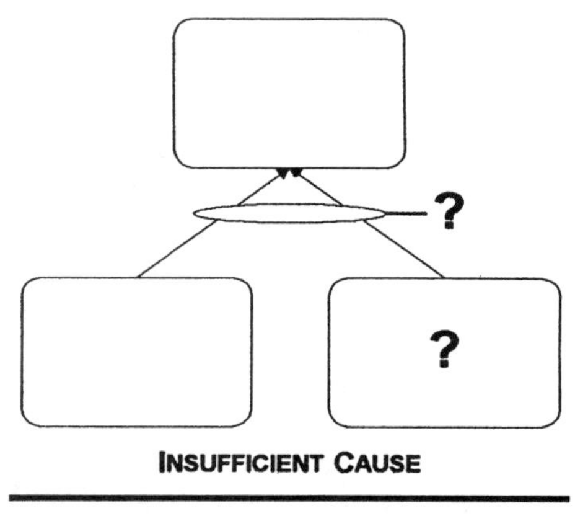

Figure 4.16

1. *Is there something else that must exist in conjunction with the speculated cause, in order for the effect to exist as a result?*
2. *Does that something exist in the reality of the system or situation that I'm examining?*

A manufacturer of temperature control devices for the injection molding industry was experiencing material shortages in manufacturing. In the process of analyzing the problem, the purchasing manager said that a cause for the material shortages was late deliveries of purchased components by their vendors.

We encouraged the purchasing manager to examine her claim a bit further. It seemed that this situation, in and of itself, could not cause material shortages to exist in manufacturing, because we knew of many manufacturing companies that did experience fluctuations in vendor performance yet did *not* suffer from manufacturing shortages. We posed a question:

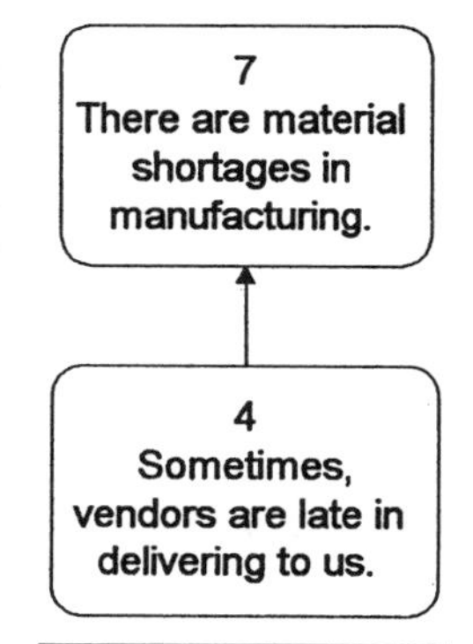

Figure 4.17

- *Did the company experience material shortages in manufacturing* ***every single time*** *a vendor was late on delivery?*

The answer to this question was no. The next thing we asked was, *What's the difference between the two situations? Under what circumstances does manufacturing experience shortages as a result of late vendor delivery, and under what circumstances does manufacturing not experience such shortages?*

In most cases, purchasing requested delivery dates from the company's vendors to coincide with manufacturing's production schedule. On some parts, the company carried an inventory in the stockroom. Manufacturing shortages on those parts were typically not due to vendor performance, but to other things (additional causes).

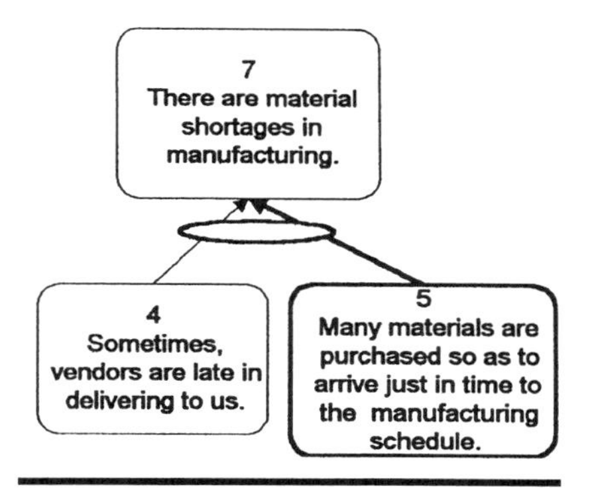

Figure 4.18

This analysis resulted in the manufacturer realizing that even though they did experience some late deliveries from their vendors, the effect — shortages in manufacturing — was a result of this *combined with their own policy* to order materials just in time. They could choose to work on their vendors, or they could choose to change their policy. Or some combination of the two. The problem wasn't just "out there in vendor land," and that was both humbling and empowering.

Like a chemical reaction, in reality it takes a combination of conditions to cause another condition to exist. Understanding the elements of those combinations can be quite important as we endeavor to make improvements in our organizations and our lives.

- We gain an understanding of the circumstances in which the cause is or is not a cause, and thus a better understanding of the circumstances in which a given solution will or will not lead to the desired outcome.
- We uncover more avenues and choices for solutions. Quite often, the newly discovered avenues are much more within our own realm of control or influence than previously believed.

After several rounds of clarity, additional cause, and insufficient cause reservations, our manufacturer's analysis of material shortages looked like Figure 4.19. Please note that Figure 4.19 is not a "pincushion." It contains three distinct and major causes for the effect.

Be careful to keep the sufficient cause diagram practical. If you find, for instance, that as a result of using the insufficient cause reservation you have more than four entities bound by an and-connector, one of two conditions exists:

- You have several layers of causality embedded in the diagram, in which case you should apply the causality reservation to clarify the cause–effect relationships.

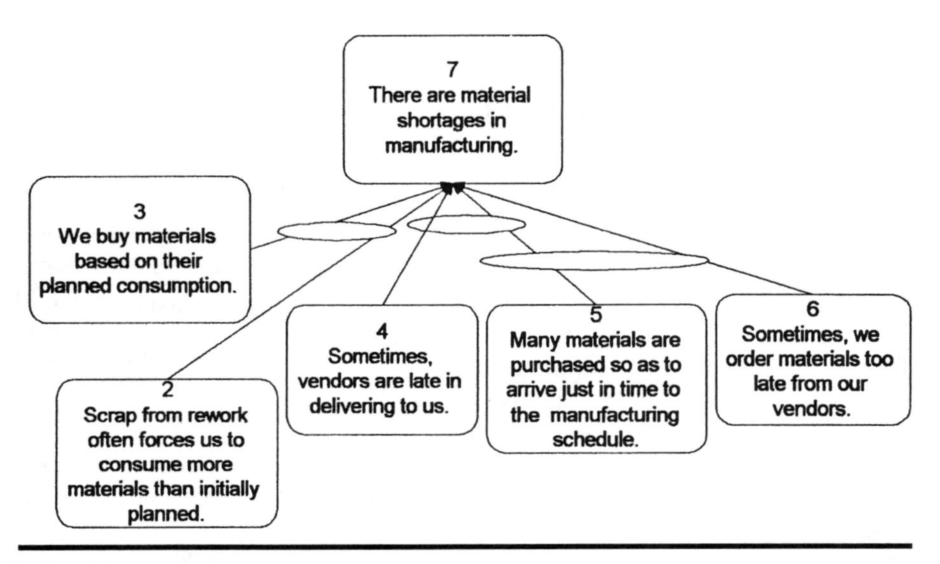

Figure 4.19

- Your diagram is overloaded with entities that are trivial — so obvious that their only purpose seems to be to clutter up the tree.

Take a look at this example, which was first presented to me by Dale Houle, Managing Partner of the Avraham Y. Goldratt Institute:

You observe a torch burning in the stadium. Why is there a torch burning in the stadium? Well, because it's the opening day of the Olympics! You are practicing your sufficient cause diagrams and draw it as Figure 4.20.

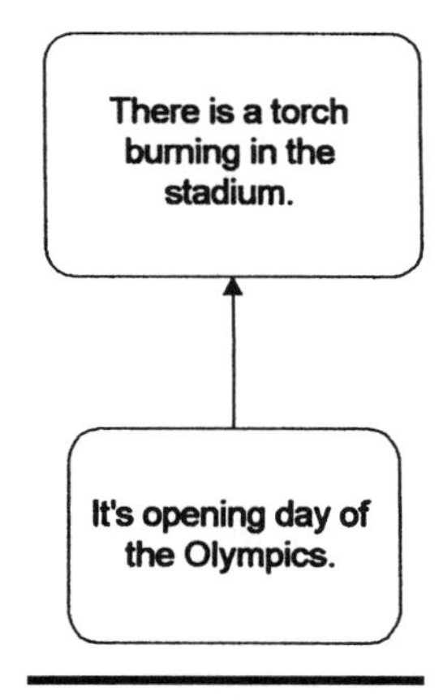

Figure 4.20

Well, my friend tells me that my diagram is incomplete. She claims that in order for my claim to be valid, there must also be *oxygen* present in the stadium. She claims that my diagram should look like Figure 4.21.

To most of us, the addition of the "oxygen" entity, although correct in reality, really adds nothing but clutter to the diagram. Now, I suppose when we ultimately discover life on other planets, and the Olympics becomes an intergalactic event, the "oxygen" entity will be more relevant.

When I say keep it practical, I mean keep it meaningful for the people who need to be involved in using (creating, reading, communicating) the diagram. If you are trying to obtain clarity of understanding, contributions to an analysis or solution, or consensus on a given subject, don't remove an entity just because it's "oxygen" to you. Doing so will send the message to your colleagues that their understanding is inferior to yours. This is not the message you want to send if you're looking for collaboration. Only remove an entity if all concerned agree that it's "oxygen."

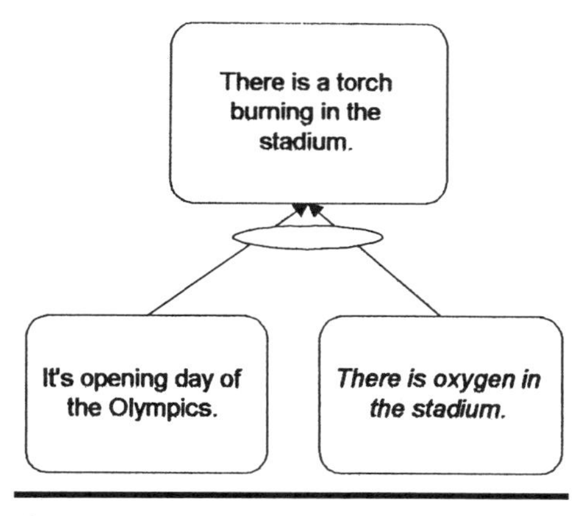

Figure 4.21

Predicted Effect

The thinking processes were founded on the *predicted effect* reservation. The rest of the reservations were discovered later, as we gained better understanding of what the predicted effect reservation was actually accomplishing. Predicted effect is nothing more than the scientific method. For a given effect, speculate its cause. Try to *in*validate your hypothesis by predicting another effect that *must* result from the cause, and check for the existence of that effect. If the predicted effect does exist in the reality of the system or situation you are examining, then your hypothesis has passed a test of validity. If the predicted effect doesn't exist in that system or situation, then you have in some way invalidated your original hypothesis. This is why the thinking processes are often called "effect–cause–effect" thinking.

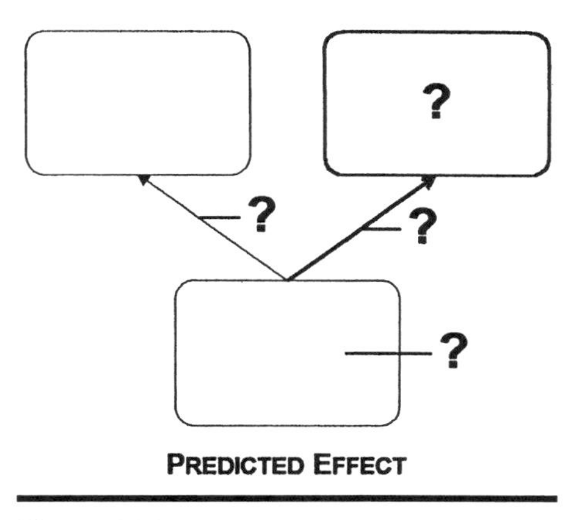

Figure 4.22

The predicted effect reservation is used to check for entity existence, especially when the entity is something intangible and difficult to verify physically. It is also used to validate cause–effect relationships.

Let's visit a nationwide distributor of electronic components who is trying to understand a substantial decrease in sales volume in spite of an increasing market for the types of products it sells. Many of the salespeople are complaining that an across-the-board price increase is the major reason for the drop in volume.

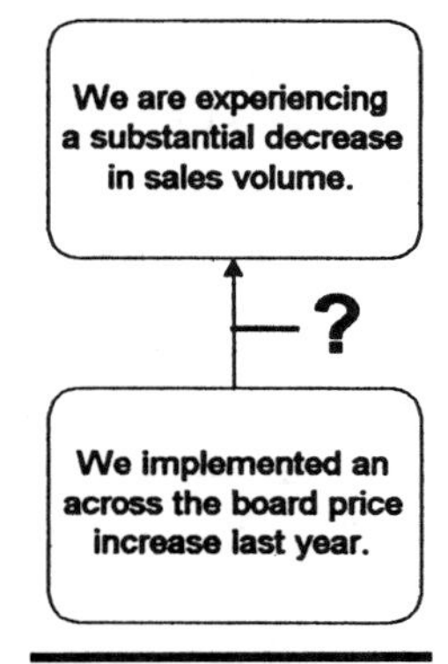

Figure 4.23

Let's say that the group generally agrees with this hypothesis. It is certainly an easy conclusion to reach, as most of the salesforce complained bitterly at the time the price increase was introduced. What would their next action likely have been? Of course, they would introduce a price cut. They would probably do it with pizzazz, announcing that they would meet or even beat their competition's prices. That might be all right, but what if the price increase wasn't the major reason for the decrease in sales? How can we check? Here's where the predicted effect reservation comes in handy.

> The great tragedy of Science — the slaying of a beautiful hypothesis by an ugly fact.
>
> Thomas Henry Huxley, 1870

The common practice would have us go about looking for all the evidence we can find to substantiate the claim that the price increase is the culprit. The predicted effect reservation takes a different approach. It asks:

- What entity, if we found it in reality, would prove our hypothesis **wrong**?

In this case, we might suggest that the company look for any of the following:

- Sales that were lost to higher priced competitors. If customers are migrating from the distributor to a higher priced source, then declining sales must be due to a reason other than price.
- Customers who are still buying products from this distributor that they can find cheaper elsewhere.

If they found either of these to exist, then their initial hypothesis would be proven incorrect. They would not be able to claim that the price increase, in and of itself, was responsible for the downturn in sales. Here's what the ensuing discussion uncovered:

In one area of the country, they were losing sales to higher priced competition. The competition was a smaller distributor who had recently hired away this distributor's best salesperson. With the salesperson went the years of relationships he had developed. His customers eagerly followed him to the other company. This accounted for 60% of the drop in sales in that area, and 15% of the drop in the company's sales.

In another area of the country, they were also losing sales to higher priced competition. It seemed that a new distributor opened up last year. At the time, they were "written off" as "nothing to worry about." A new upstart with crazy ideas. They offered customized services, such as daily replenishment, no minimum order, dedicated customer service people, and 24-hour access. This type of service was almost unheard of in this area, and most distributorships didn't believe customers would pay for it, nor did they believe that doing business in this way could be profitable. This situation accounted for about 50% of the drop in sales in that area, and 30% of the drop in the company's sales.

Both of the above scenarios pointed to the relationships the company maintained with its customers rather than its pricing strategy.

Finally, the company examined some of its customers who, in spite of the fact that they could buy products cheaper from other sources, chose to stay with the company. Generally, these customers chose to stay because of the rapport they had established with the distributor over the years. They felt that they could count on good service, they believed that the salesperson they worked with had their best interests at heart, and they knew they could count on their salesperson to come through for them when needed.

So it seemed that in many of the cases, customers left or stayed not because of price, but because of relationship. After all of that, the diagram now looked like Figure 4.24.

Using the predicted effect relationship helped to uncover additional, important aspects of the cause–effect relationship that were not evident before. The company now has the opportunity to work at market segmentation in a way that preserves and enhances their customers' perception of value, and subsequently the prices they pay. They have an opportunity to work on developing strong relationships and new services that will foster that increased perception of value. If they choose, they can carefully select products and/or markets in which to play the price game.

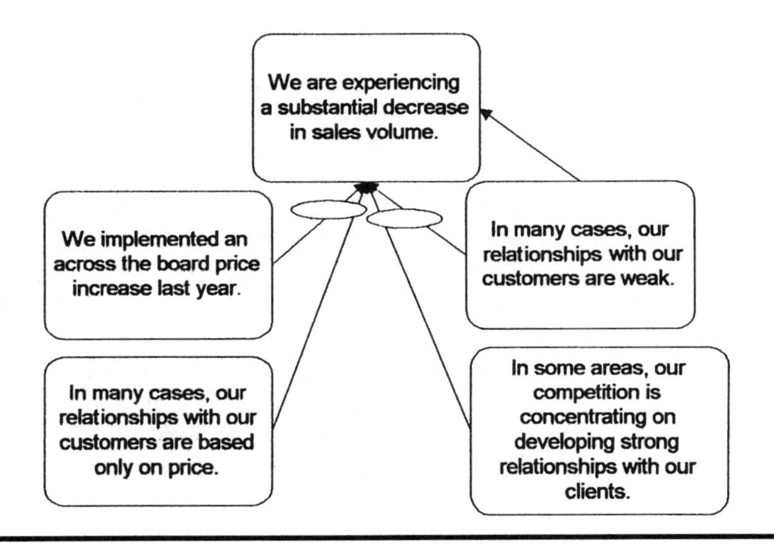

Figure 4.24

Reservations Disguised as Common Expressions

We use these reservations every day, but often don't realize it. Do you remember the car battery example? As soon as we started to check the lights and the radio, we were using predicted effect.

Here are ways in which reservations are expressed in our day-to-day conversation:

Oh, really?!
No way!
Yeah, but...
Bull----!
I don't think so...
No, I think it's [fill in the blank].

Some Guidelines

When you are on the receiving end of communication, use the clarity reservation first. It shows that before you assume the other person is incorrect, you first assume that you might not fully understand what he or she is trying to communicate.

When you are creating sufficient cause diagrams, don't worry about using the clarity reservation first, but remember that you are not ready to show your diagram to anybody else until you have applied the clarity reservation to it.

When you are ready to share a diagram with others, add whatever is necessary to make it clear. Don't leave entities off because you consider them to be "oxygen" if someone involved in the diagram believes the entity to be more relevant than that.

Don't forget to put on your learning hat, and remember that you are trying to prove yourself wrong every step of the way. This is an about-face from our common practice of "being right" and trying to substantiate our claims every step of the way. I hope that you have seen a glimpse of the benefits of this type of outlook — you empower yourself to learn and to recognize opportunities that you were previously blind to.

Skill Builder: Using the Categories of Legitimate Reservation

The more often you do the following exercise, the more proficient you will become in the use of the categories of legitimate reservation, sufficient cause thinking, and communicating.

1. Think of a sufficient cause relationship and write it down.
2. Give a copy of this to a friend or colleague and ask them to write down any questions or disagreements they may have with the statement.
3. Create a sufficient cause diagram of the relationship you wrote down in step one.
4. For each question or disagreement that your friend wrote, identify the reservation category that he or she is expressing.
5. Make any modifications to the drawing necessary to resolve each reservation (by adding and/or subtracting entities and arrows) (see Figures 4.25 and 4.26).
6. Continue this process until your friend says something like, "I have no more questions — this is crystal clear, makes sense, and I believe it!"

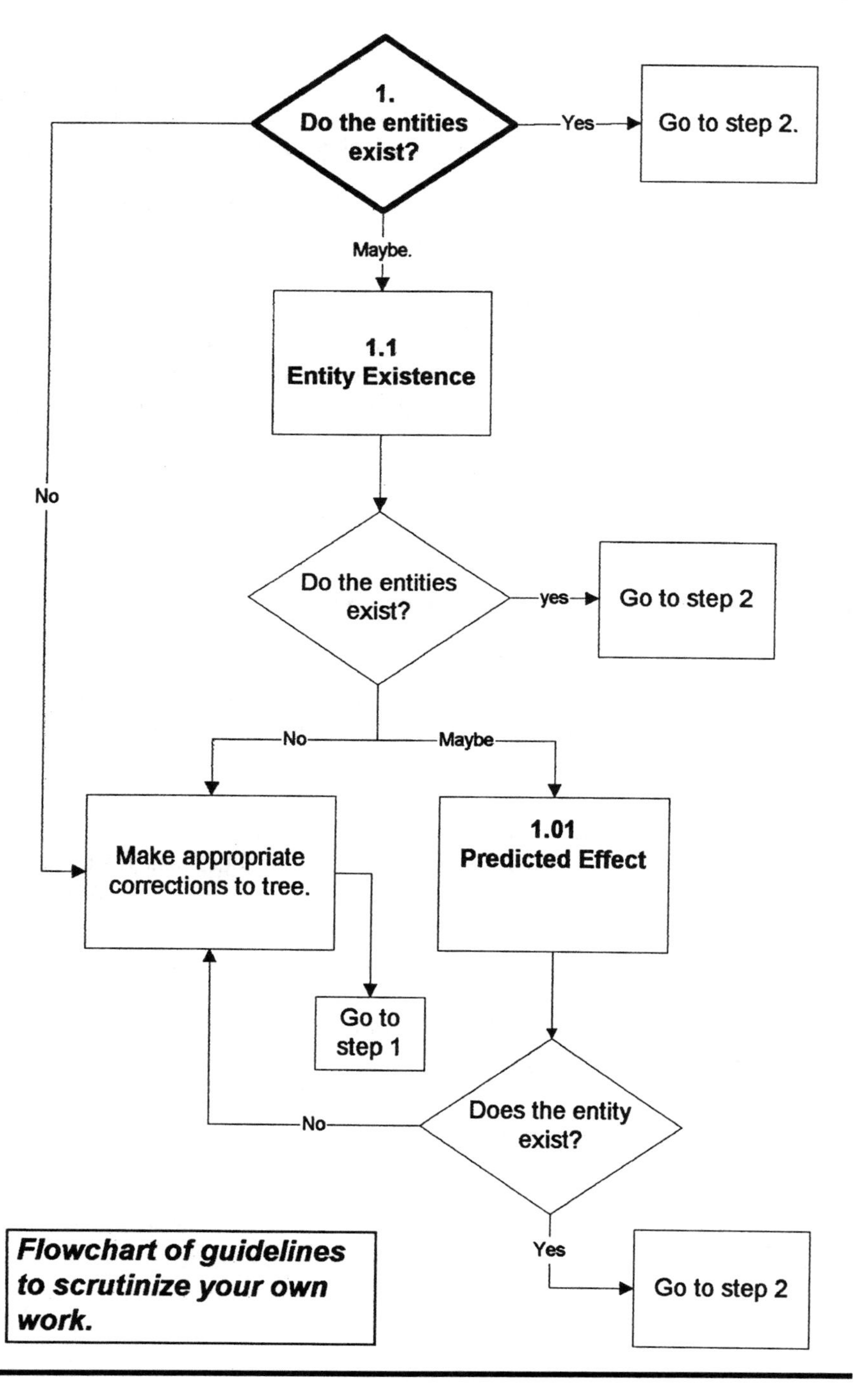
1.
Do the entities exist?
Yes
Go to step 2.
Maybe.
1.1
Entity Existence
No
Do the entities exist?
yes
Go to step 2
No
Maybe
Make appropriate corrections to tree.
1.01
Predicted Effect
Go to step 1
No
Does the entity exist?
Yes
Go to step 2
Flowchart of guidelines to scrutinize your own work.

Figure 4.25

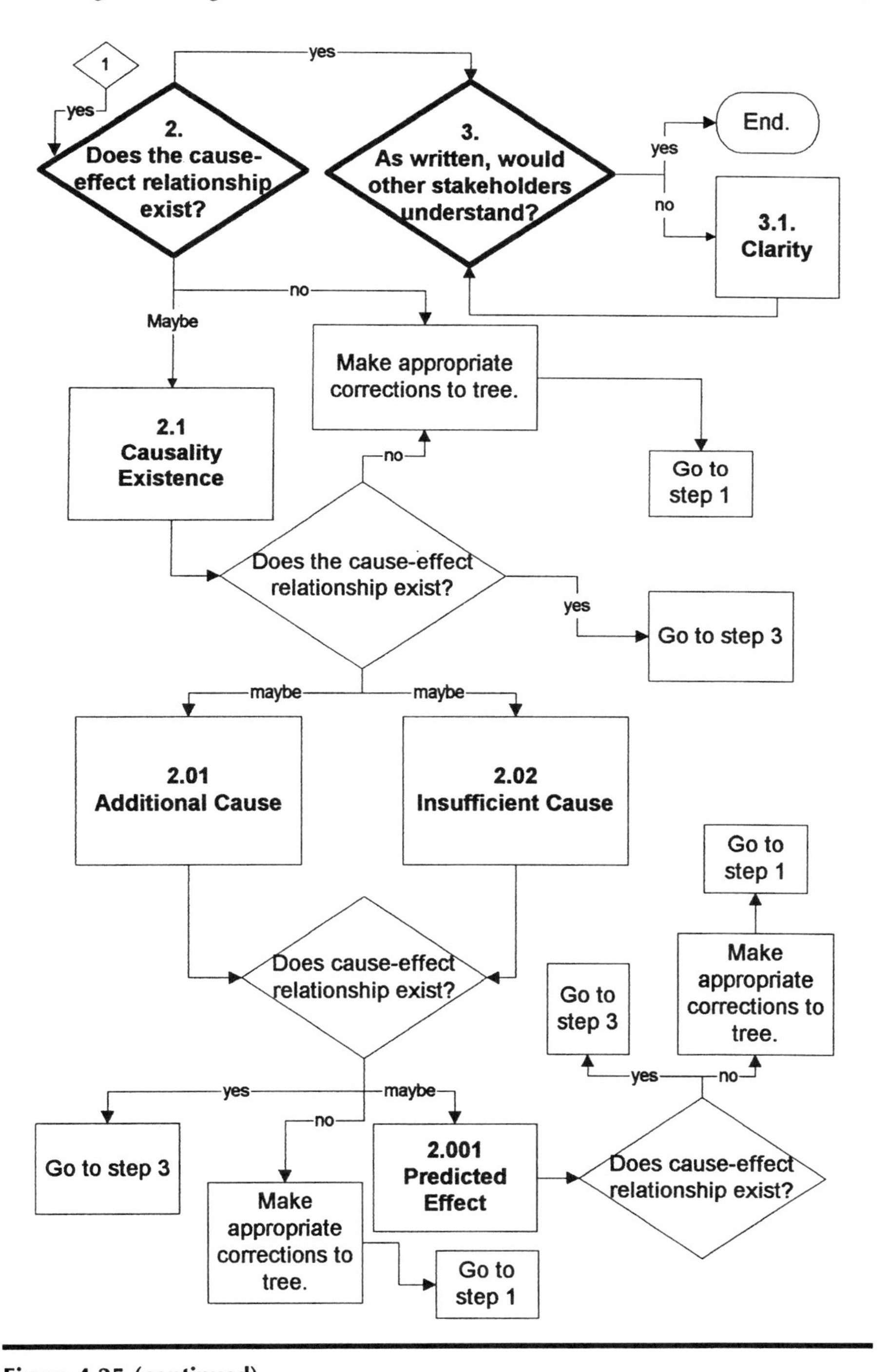

Figure 4.25 (continued)

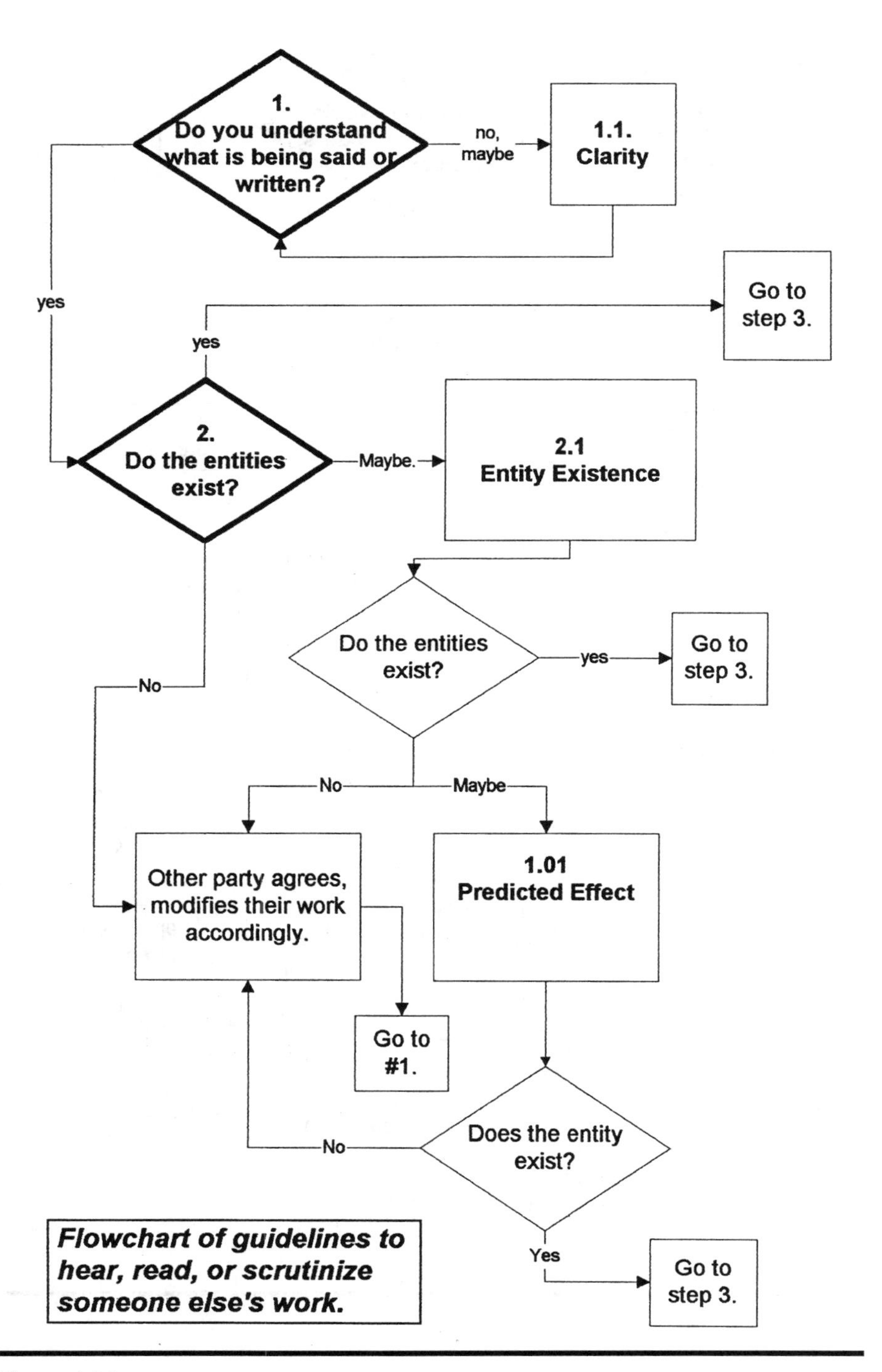
1.
Do you understand what is being said or written?
no, maybe
1.1.
Clarity
yes
yes
Go to step 3.
2.
Do the entities exist?
Maybe.
2.1
Entity Existence
Do the entities exist?
yes
Go to step 3.
No
No
Maybe
Other party agrees, modifies their work accordingly.
1.01
Predicted Effect
Go to #1.
Does the entity exist?
No
Yes
Go to step 3.
Flowchart of guidelines to hear, read, or scrutinize someone else's work.

Figure 4.26

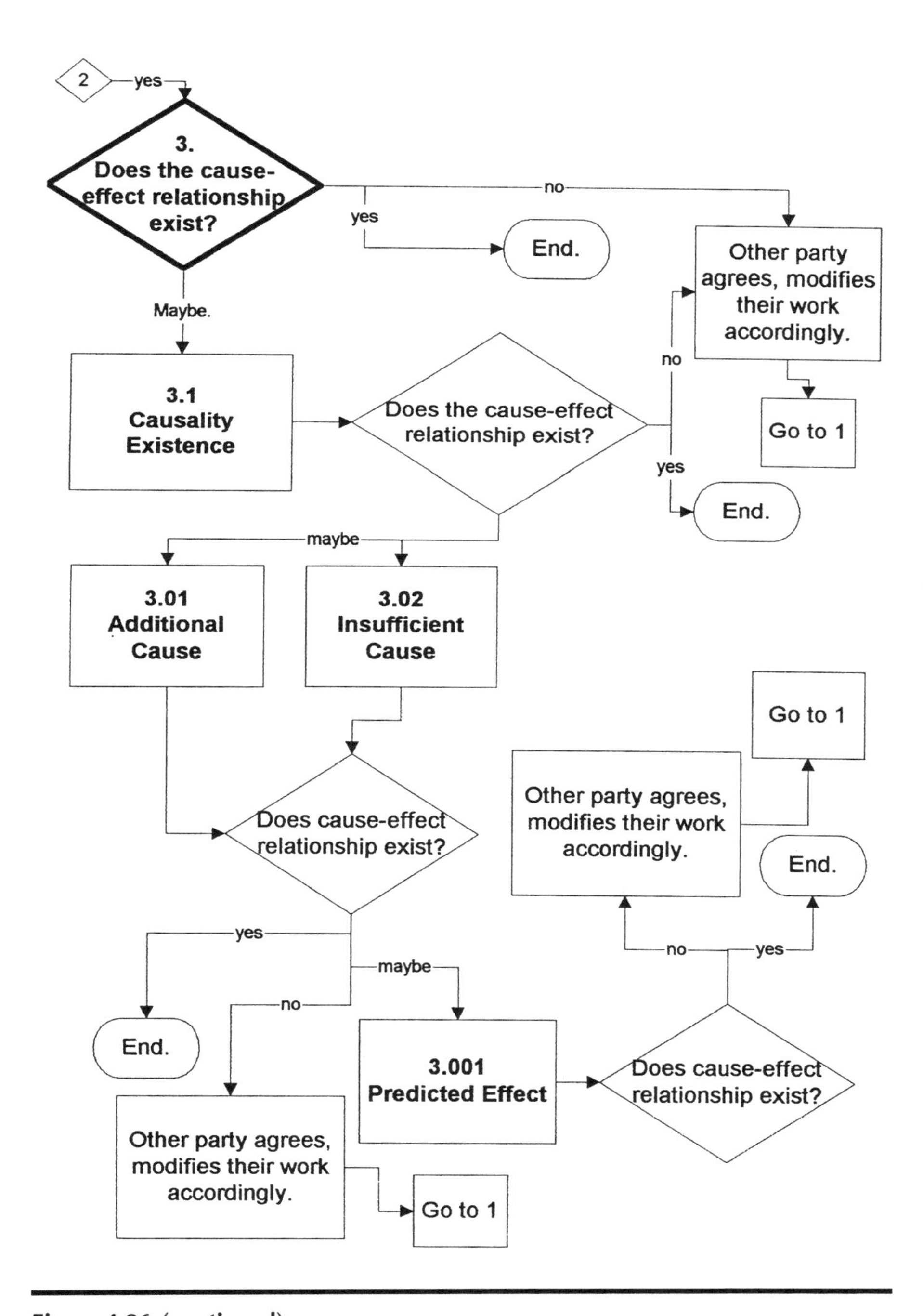

Figure 4.26 (continued)

Chapter 5

Necessary Condition Thinking

Necessity is an evil; but there is no necessity for continuing to live subject to necessity.

Epicurus, 3rd Century B.C.

Necessary Condition is the thought pattern we use when we are thinking in terms of *requirements*. When we think that something *must* exist before we are able to achieve something else, we are using necessary condition thinking. Terms such as *must, must not, cannot, need,* and *have* are indicators of necessary condition thinking.

Necessary conditions are rules, policies, or laws, that provide the limitations, or boundaries within which we believe we are allowed to pursue goals and objectives. Conformance with a necessary condition does not guarantee that a goal will be achieved, but we usually believe that if we don't have the perceived necessary condition in place, we will certainly be unable to attain that goal.

Here are some examples of necessary condition thinking:

- *A person must be at least 35 years of age to be the President of the United States.*
- *I can't be physically fit unless I exercise.*
- *We must reduce our costs in order to get higher profits.*

Our interpretations of necessary condition relationships shape our perceptions of what we are and are not able to do. Often, we limit our opportunities by believing strongly in necessary conditions that aren't, or don't need to be. As a result, we block ourselves from seeing and acting upon many simple, practical, creative, and speedy ways to achieve our objectives.

A Simple Example

The following example (Figure 5.1) can be found in many texts and workshops on creative thinking, and you may have experienced it before. Without lifting your writing instrument, and without retracing any lines, connect all nine dots by drawing four straight lines. If you have difficulty accomplishing this task, you are assuming a necessary condition that doesn't exist. If you are familiar with this puzzle, try it with three lines, two lines, or just one line. Hints and answers are provided in the Appendix. Please, try the exercise before you look! A hint: What are you assuming about the instructions that is not explicitly stated in the instructions? Yes, it is possible!

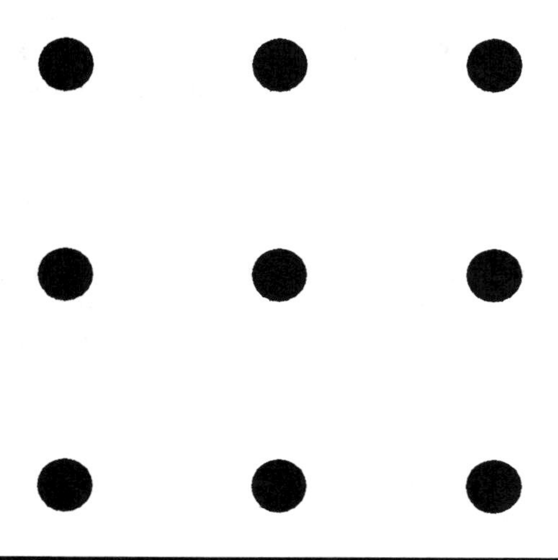

Figure 5.1

Defining, Understanding, and Changing Necessary Condition Relationships

The Thinking Processes provide systematic, three-step processes to challenge perceived necessary conditions through uncovering and challenging the assumptions that formulate the necessary condition relationship.

1. Diagram the necessary condition relationship.
2. Surface the underlying assumptions.
3. Brainstorm alternatives.

Diagram the Necessary Condition Relationship

Every necessary condition diagram is comprised of the following elements:

1. Entity (Figure 5.2)
2. Arrow (Figure 5.3)
3. Necessary condition (Figure 5.4)
4. Objective (Figure 5.5)
5. Assumption (Figure 5.6)

Entity

An entity is a single element of the system. Although the sufficient cause processes require that entities be written as complete sentences, it is not a requirement in necessary condition diagrams.

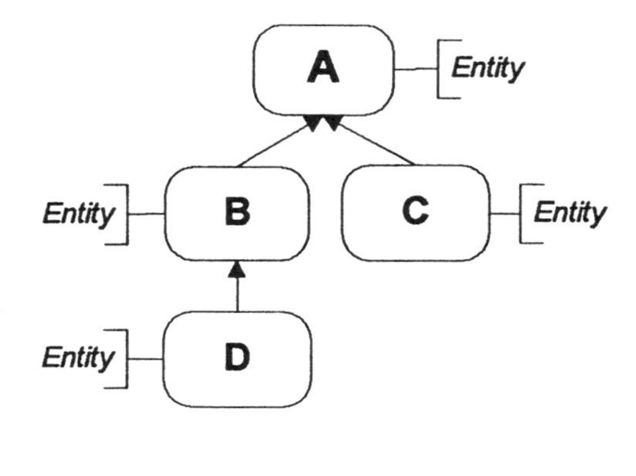

Figure 5.2

Arrow

An arrow indicates a relationship between two entities. The entity at the base of the arrow is the *necessary condition* (see Figure 5.4). The entity at the point of the arrow is the *objective* (see Figure 5.3).

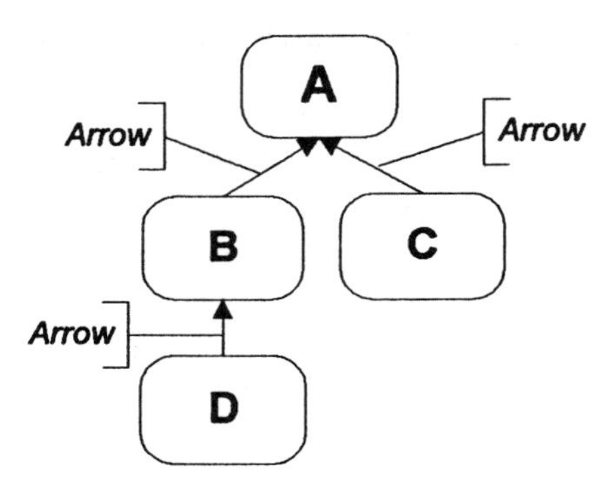

Figure 5.3

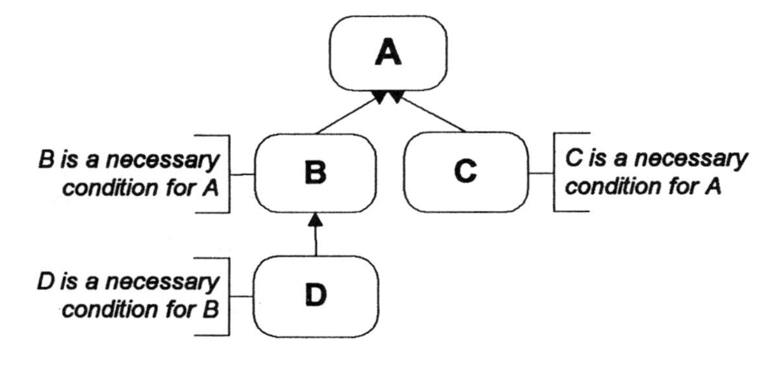

Figure 5.4

Necessary Condition

An entity is a *necessary condition* when it is considered to be required in order for another entity (the objective) to exist or to be allowed to exist. The necessary condition is located at the base of an arrow (see Figure 5.4).

Objective

An entity is an *objective* when it is thought that the entity cannot exist unless another entity, the necessary condition, exists as well. The objective is located at the point of an arrow (see Figure 5.5).

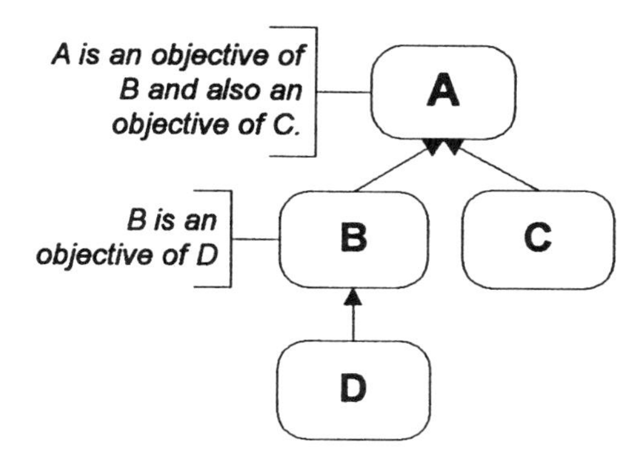

Figure 5.5

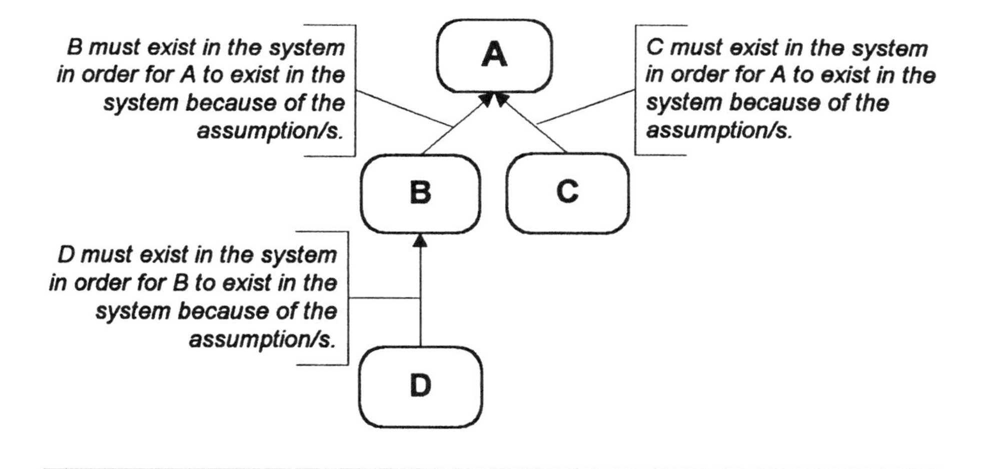

Figure 5.6

Assumption

An assumption is a reason for the existence of the necessary condition relationship — a reason that one entity is required in order for another to exist. You might say that the arrow represents the "space" where the assumptions hide. When somebody says, "what's underneath the arrow," they are really asking you to explain the assumption that connects the necessary condition to the objective. Assumptions are valid or invalid (see Figure 5.6).

Reading the Necessary Condition Diagram (Before Assumptions)

You can check a necessary condition diagram by reading it back to yourself. Any of the following statements should flow smoothly when you have filled in the blanks with the relevant entity:

- In order to achieve [objective], I must [necessary condition].
- Before we can accomplish [objective], we must [necessary condition].
- We can't get [objective] without [necessary condition].
- You have to [necessary condition] before you'll be able to [objective].
- Unless we have [necessary condition], it will be impossible to achieve [objective].

Reading the Necessary Condition Diagram (Including Assumptions)

Any of the following statements should flow smoothly when you have filled in the blanks with the relevant entity:

- In order to achieve [objective], I must [necessary condition] because [assumption].
- Before we can accomplish [objective], we must [necessary condition], because [assumption].
- Because of [assumption], we can't get [objective] without [necessary condition].
- Due to [assumption], you have to [necessary condition] before you'll be able to [objective].
- Unless we have [necessary condition], it will be impossible to achieve [objective], because [assumption].

Examples

Using the nomenclature described above, the examples of necessary condition thinking described at the beginning of this chapter can be diagrammed as in Figure 5.7.

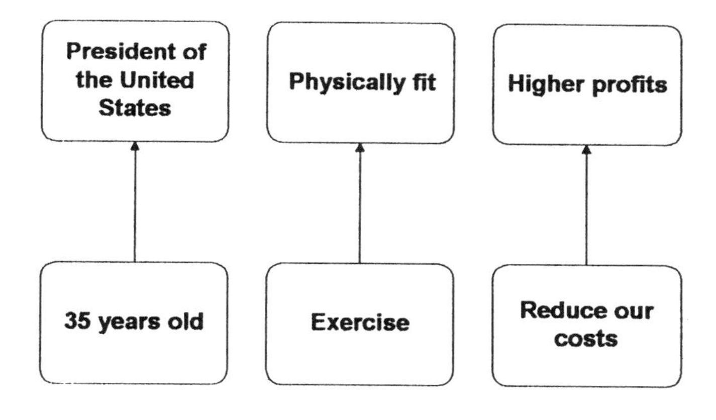

Figure 5.7

Surface the Hidden Assumptions

Let's examine one of the examples: *"We must reduce our costs in order to get higher profits."* In the 1980s and the first half of the 1990s, this seemed to be the cardinal rule of industry. In fact, many of you might even say that it is still the main focus of your companies. The most sought-after executives were those who "trimmed the fat" by "right sizing" their companies into lower cost structures. Many of these companies learned the hard way that this approach led to very short-term improvements in profitability, and long-term deterioration of morale, productivity, and profitability. Let's see if we can uncover the key assumptions that led us to believe that this was *the* way to improve the profitability of our enterprises, and then challenge ourselves to find alternatives to this methodology. We do this by asking questions that point directly to the assumptions that lurk beneath the arrow.

- *Why* do we believe that we must reduce our costs in order to get higher profits?
- *Why* do we believe that we cannot get higher profits unless we reduce our costs?

Our answers to these questions uncover some of the key assumptions that were used to form the necessary condition relationship between cost reduction and higher profits. In this example, the assumptions might include:

- *Because* our ability to sell more volume is limited. (*In order to achieve higher profits, we must reduce our costs, because our ability to sell more volume is limited.*)
- *Because* our products are already selling at the maximum price the market will pay. (*The reason we must reduce our costs in order to achieve higher profits is that our products are already selling at the maximum price the market will pay.*)
- *Because* there is a tremendous amount of fat in our system that cannot be put to productive use. (*We must reduce our costs in order to achieve higher profits because there is a tremendous amount of fat in our system that cannot be put to productive use.*)

When assumptions are visible, we can test both their validity and necessity.

- Are we *really* selling our products and/or services for the highest price possible? Perhaps there are market segments that would place a higher value on what we sell. Have we looked for such a segment?
- Is our market *really* limited? Have we ignored capabilities of our organization that would be of value in the marketplace?
- Do we *really* have too much fat in the system? If we trimmed the so-called fat, will we limit our ability to be flexible and responsive to our markets?

Brainstorm Alternatives

Now that we have uncovered and challenged several assumptions, we may find one or more to be invalid. We may also find one or more assumptions that we'd like to cause to become invalid. In any case, we have a tremendous opportunity. The previous boundaries within which we attempted to achieve our objective have been broadened, and we are free to explore new solutions. The next step, therefore, is to brainstorm alternative solutions. We refer to these alternatives as "injections." If we "inject" the alternative solution into the environment, the necessary condition would no longer be required in order to achieve the objective. Finding injections means asking more questions:

- How can we achieve our objective *without* the necessary condition attached to it?
- What, if we implemented it, would enable us to render the key assumptions invalid or irrelevant?

> Just as our eyes need light in order to see, our minds need ideas in order to conceive.
>
> Nicolas Malegranche, 1674

Let's try this approach in our quest to unlock the paradigm of cost reduction as a necessary condition to higher profits. List as many ideas you can.

- How can we achieve higher profits without reducing costs?
- What, if we implemented it, would enable us to sell more volume? Unlimited volume?
- What, if we implemented it, would enable us to sell our products for higher prices?
- What, if we implemented it, would result in a highly productive system? What, if we implemented it, would transform the fat into muscle?

As you can see, focusing on the assumptions — what are they and how might we make them invalid — provides us with a key to opportunities to which we were previously blinded.

The Reference Environment Method: Getting Unstuck

What if you are unable to surface the assumptions? What do you do if your only answers to the questions designed to uncover the assumptions are "Well, because!" or "I don't know," or "Because that's just the way it is!"

What do you do when you find yourself coming up with the same old solutions (that haven't worked yet) as you attempt to brainstorm alternatives? What if you can't generate *any* alternatives?

Use your life experience and imagination to come up with analogies, or reference environments. Try to think of an environment, or a situation, in which the objective or something like it exists and the necessary condition does not. Once you have that environment or situation in mind, ask yourself the following questions:

- Why can that environment have the objective without the necessary condition?
- Why doesn't that environment need the necessary condition to achieve the objective?

Use your answers to surface the assumptions of the necessary condition relationship that you are trying to understand and challenge.

- What does this tell me about the necessary condition relationship that you are examining?
- What assumptions are missing from that environment that are present in yours?
- What does that environment have that your environment doesn't?

Let's try this method with another of our examples. Let's see if we can generate new ways to become physically fit without the hard work of including exercise in our fitness regimen. Can you imagine any environment in which something analogous to physical fitness is achieved without exercising? Here are some that I've imagined:

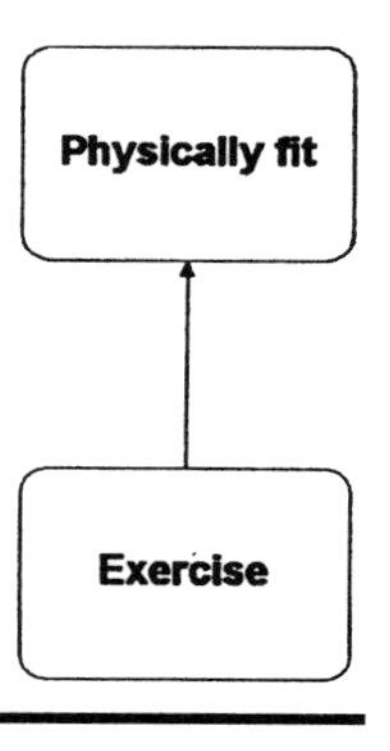

Figure 5.8

An animal, such as a deer, in the forest. I don't picture a deer working out on a treadmill, or grimacing while lifting weights, yet most deer I've seen seem pretty fit. They are lean, fast, and graceful. *Why?* What's different about the deer's environment?

- Most deer don't have access to junk food. The forest provides them with the nutrition they need.
- Deer seem to be preprogrammed to wander around and run in the forest. Thus, they get the exercise they need simply by living and doing what they're naturally programmed to do.

Asking, *what do these differences tell me about exercise as a necessary condition to fitness?* led me to the following: I need to exercise in order to become physically fit because my life-style does not naturally provide the exercise my body needs. Understanding this assumption helps me brainstorm more alternatives, such as these that some of my students have suggested:

- Get a job that incorporates physical activity, such as postal carrier, construction worker, or bicycle courier.
- Join the military.
- Move to the forest.

Subliminal learning. Thinking about animals that don't need to work at exercising in order to be fit led me to thinking in general about achieving goals without working at them. This led me to the subject of subliminal learning — the ads say that you can simply play the tape while you're asleep and learn a new language! *Why can some people learn a new language while they sleep, and I can't become physically fit if I don't exercise?*

- Because you must be awake to exercise your body
- Because you can't sleep while you're exercising your body

The assumption that this train of thought led to was, *We must exercise in order to become physically fit, because exercise is the only way in which we can put our bodies to work.* Utilizing the subliminal learning analogy, students have offered the following injections:

- Develop a machine that exercises you while you sleep.
- Develop a medication to stimulate the nervous system and muscles so the brain will think the body is exercising.
- While exercising, read or listen to something so mentally engrossing that you're almost unaware of the fact that you're exercising.
- Meditate while exercising, so that you're not aware that you're exercising.

The purpose of the examples is not to get into a debate about the practicality of any of the ideas, or even the exactness (or lack thereof) of the analogies. Rather, I share them to illustrate this method as a means for opening up your thinking and unlocking some of your paradigms. I want to encourage you to be bold and creative as you use this approach. My experience is that out of bold, creative ideas, come quite practical and usable ideas. Just as professional photographers often shoot many rolls of film in order to get just a few great shots, it is helpful for us to generate lots and lots of ideas, in order to find the one or the few that we actually want to implement. Have fun with it! Allow yourself to go on the thinking journey to generate as many ideas as you possibly can. Don't block yourself from writing down an idea just because it seems impractical. Writing it down frees you to generate and write down the next one, and the one after that, and the one after that.

Skill Builder: For the Next Couple of Weeks, Be Aware of Your Use of Necessary Condition Thinking

- Choose one instance each day (from your conversations, from your newspaper, TV news program, or other sources), and diagram it as you've seen in this chapter.
- Write down the key assumptions that form the bond between the objective and its perceived necessary condition.
- Brainstorm ideas which, if implemented, would negate the key assumptions. Be creative!
- If you hesitate to write down an idea because *it's crazy* or because it's *not implementable*, write it down anyway. And then look for the assumptions that caused you to believe so!

Part Two

Chapter 6

Transition Tree

> A thought which does not result in an action is nothing much, and an action which does not proceed from a thought is nothing at all.
>
> Georges Bernanos, 1955

This afternoon, I observed a group of middle managers talk about changing the culture of their plant. They began their discussion with a decision to hold a series of small group meetings with the employees and then began to plan the agenda for those meetings. *"We'll start by letting them know what the productivity trend is, and that the plant won't last if the trend continues... We'll let them know that labor is the largest component of OE (operating expense), and show them how much of the OE is overtime. We'll tell them that we don't have time to let the changes just happen little by little, that we need to make it happen now..."* I found their discussion fascinating, because I knew that they had not spent any time before this session to clarify what the plant's culture was going to be once it was changed. They had not spent any time determining the specific changes in behavior they expected to see (their own as well as the employees'). And, even though they were deciding what they were going to say in these small group meetings, they had not spent any time contemplating what they expected any of these small groups of people to *do* once the meetings were over. At this point, what do you think were the odds that the small group meetings would bring their plant closer to a changed, improved culture?

The group's facilitator finally spoke up. Suggesting that they were putting the cart before the horse, he challenged them to clarify what they wanted to happen as a result of these meetings before working on the content and flow of the meetings. They spent the rest of their time doing just that. Tomorrow, they will use the transition tree to create the agenda for the small group meetings that will be but one part of the process to create and sustain improvement in both the plant's culture and productivity.

We live in a very action-oriented society. We don't feel productive unless we're *doing* something. Our plans, agendas, outlines, and process flow diagrams are filled with "do" items. *First we'll do this, then we'll do that, and then we'll do item number three on the list.* What's missing from this equation? We forget that for every action there's at least one reaction. We forget that, when we have a plan that contains steps that must be performed in a specific sequence, there are reasons for sequencing the steps. In other words, each step is (or at least should be) an action that is meant to create a reaction — a new condition. Let's call this new condition C^1 (Figure 6.1). By combining condition C^1 with the next planned action, yet another new condition (C^2) is created. Of course, condition C^2 is meant to bring us closer to the objective of the plan. Until condition C^1 exists in reality, that next step is a waste of time. We typically go wrong in two aspects. We don't verbalize the objectives of our actions (let alone our action plans), and we don't clarify what we expect to happen from each step. Either of these mistakes will give us pretty good odds of either not achieving our objectives, or of taking more time than necessary to achieve our objectives.

The Process

The transition tree is a sufficient cause diagram used for creating action plans. The transition tree contains four types of entities, as illustrated in Figure 6.2:

A. The injections are **actions**. These are the specific things that are to be done in order to carry out the plan.
B. Entities that exist in the present reality are always entries to the tree. The current situation should be taken into account when developing any action plan.
C. Entities that will exist in the future are the results (effects) of the combination of implementing the actions *and* the presence of the current and future conditions that are captured with it by and-connectors.
D. The **objectives** of the action plan are achieved as a result of the conditions created by implementing the actions.

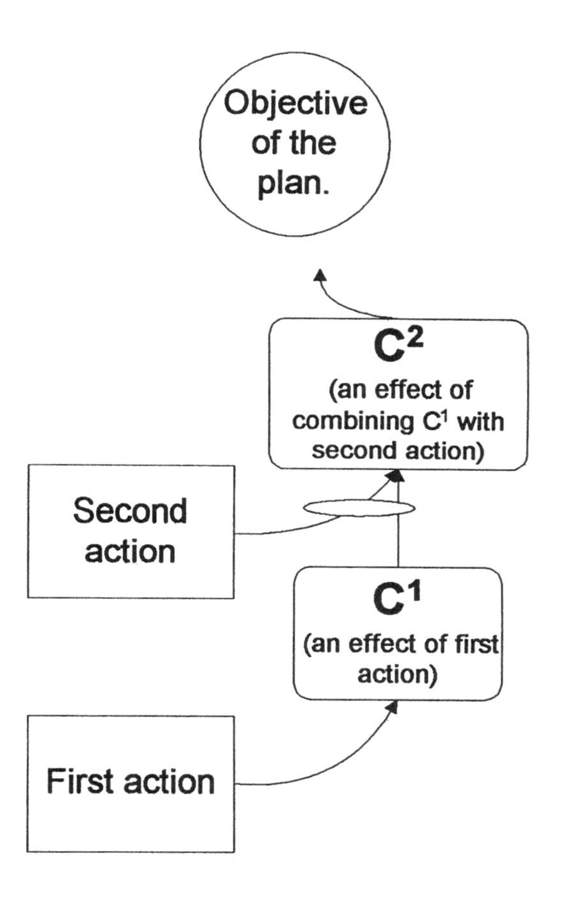

Figure 6.1

Envision a manufacturing process. Every manufacturing process starts with at least one material that has been purchased from an outside source. Every step of the way, that material goes through transitions — a resource (person and/or machine) does something specific to that material, changing it to a state that is closer to its finished form. From that state, the next resource performs a specific step in the process that will change it a bit more, moving it still closer to its finished form. And so on until the company has a finished, salable product. You will use the transition tree to design the *process* that will create the necessary transitions from the conditions present in the current reality to the different conditions you desire at some point in the future. With the transition tree, you will define the specific steps (actions) that will transform your current reality (raw materials) into a specific future reality (objectives). You will also verbalize

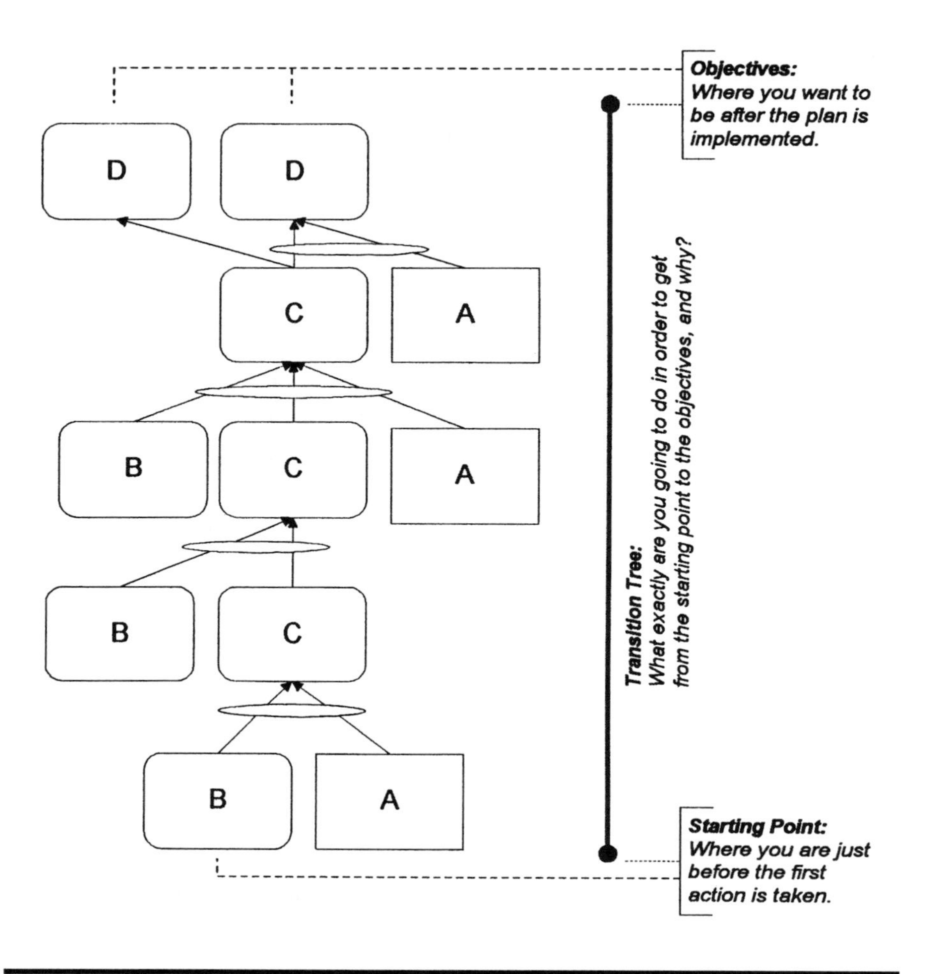

Figure 6.2

what the transitions themselves will be — the intermediate states that will be created along the way.

The transition tree is the tool to use when you need to create an action or implementation plan, and you already have some ideas in mind relative to what you're actually going to do. The objective is something you know is within reach. You may even have already determined some of the actions that you are going to take in order to reach the objective, yet it's not a "no brainer" objective. You sense that it would be really beneficial to sit down and do some planning so you can be confident that by executing the plan, the objectives will be achieved in a way that doesn't create unwanted side effects (Figure 6.3). Of course, there are times when

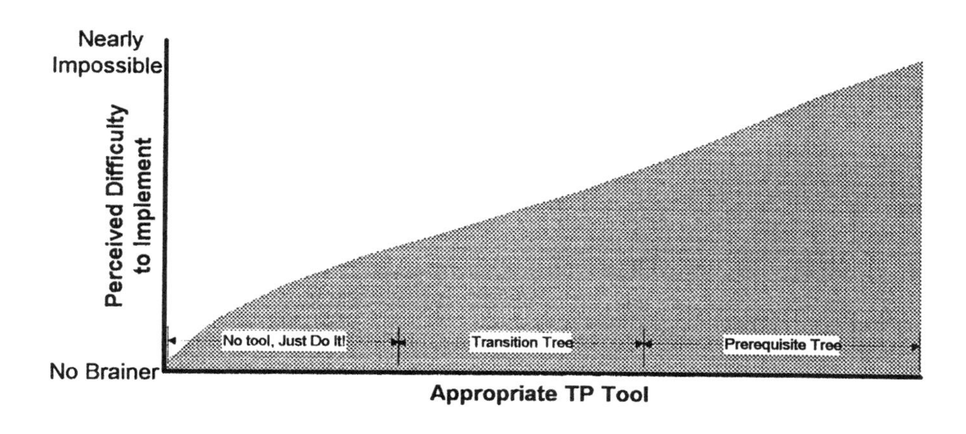

Figure 6.3

we are assigned a task or have an objective that we want or need to achieve and we simply have no clue on just what to do to achieve it. In such cases, use the prerequisite tree (see Chapter 10) before the transition tree. The prerequisite tree will guide you through verbalizing and addressing the obstacles to reaching your objective and turn what seems almost impossible into a series of reachable milestones.

Some situations in which I use transition trees are when I am

- Designing the flow of a speech, seminar, or workshop
- Planning a meeting with a client or a prospect
- Preplanning a meeting, telephone call, or conversation for emotionally charged issues
- Deciding on the specific actions that will be taken to implement strategic plans

The general steps of the transition tree process are:

1. Establish the scope of the transition tree.
2. Using sufficient cause thinking, link the initial action to the objectives.
3. Seek and block undesirable consequences.
4. Implement the plan!

After I go through each of these steps in detail, I will provide some transition tree examples. Do you have any plans that you need to make right now? Of those, are there any you should think through before you actually start acting on them? Pick one, and do the tree while we go through the steps that follow.

First, here are two general guidelines.

- **Verbalize the entities in present tense terms.** Stating the entities as if they exist in the present accomplishes a few things. It helps you to project yourself mentally into the future and better visualize what you're thinking about. Second, when most of the entities on a tree contain the word "will" (as in *we will be more profitable*), the tree tends to be more cluttered and confusing for both the tree's creator and its readers.
- **Do not use tentative types of phrasing such as maybe or possibly** (as in *we might be more profitable*) in the future reality tree. As soon as you feel yourself reaching for the M-word ("maybe"), recognize that you have a causality reservation. Resolve the reservation by utilizing the categories for legitimate reservation. This will surface the clarification you need about what else must exist in order to make your vision a reality and lead you toward adding injections and clarifying your assumptions about the environment that will make it so. Don't settle for tentativeness.

1. **Establish the scope of the transition tree.** In this step, you will define the starting elements of your transition tree.
   a. **Ascertain the premise for the transition tree.** What are you planning? Take a moment to write down (or at least say to yourself) what you are about to plan. Is it a meeting, a phone call, a course, a lecture, a sermon? Are you creating a new business process or planning the implementation of a project? This helps you to focus on the task at hand. I find statements such as the following to be quite helpful:
      - *This transition tree is to be the agenda for my upcoming meeting with John Smith.*
      - *This transition tree is to be the detailed plan of the two-hour presentation on TOC that I am to give at the next IMA conference.*
      - *This transition tree will define how I will prepare for the upcoming marathon.*
      - *This transition tree will describe the content and flow of the proposal we will submit to ABC Corporation.*
      - *This transition tree will define the process by which our customer service associates handle requests for quotation.*
      - *This transition tree will describe exactly how we are going to achieve the first milestone of our ERP system implementation.*

b. **Define the objective/s of the transition tree.** Once you have made the decision to create a plan, do what Stephen Covey says, and *begin with the end in mind.** What is the purpose of the plan? What do you expect to accomplish? Once you've done all the things your plan tells you to do, what should happen as a result? Referring back to the manufacturing process analogy, what is the finished product? Write the objectives of the transition tree as entities, using present tense.

- When I design the content and flow of a workshop, this is the place that I articulate the learning objectives. In order to do so, I formulate answers to questions such as:
  - *Upon completion of this workshop,*
    - *What will the students know?* ("Students know when to use a transition tree.")
    - *What will they be able to do?* ("Students are able to create a transition tree.")
    - *What will they want to do?* ("Students want to use the transition tree in the future.")
- When I use the transition tree to determine the agenda of a meeting, the objectives are the answers to the following questions: *What is the purpose of this meeting? What will be accomplished in it? What do I expect the participants to actually do, think, and/or feel when the meeting is over?* For instance, the managers of the manufacturing plant that I discussed at the beginning of this chapter spent some time deciding specifically what they hoped to accomplish in the small group meetings. Two of their objectives were:
  - Employees understand that the plant's productivity (T:OE) is on an unacceptable trend.
  - Operators agree to set up and run product toward the end of the shift, even if that means that the next shift finishes the batch.

A colleague recently vacationed at a large, well known, very expensive resort. The service was, as he put it, "worse than awful." He decided to write a letter to the hotel management to let them know how angry he was with the service. Before he wrote the letter, he decided to do a transition tree to determine the content and flow of the letter. His initial thinking was, "I'm going to write this letter to let them know how angry I am." When he decided to do the transition tree, he

* Covey, Stephen R., *The 7 Habits of Highly Effective People,* Simon and Schuster, 1989.

asked himself, "What do I want the resort management *to do* once they've read my letter?" He decided that what he wanted them to do was to apologize and not charge him for his stay. The result of his letter was not a free stay, but the management did credit his account for $500. He believes that, had he not done the transition tree first, he wouldn't have received anything.

- The objectives for the transition tree are often provided by one of two other thinking process application tools:
  - Injections from a future reality tree.
  - Intermediate objectives from a prerequisite tree.
- Write the objectives of the transition tree as entities, using present tense.

c. **Verbalize your starting point.** This step asks you to think about, and jot down, the environment from which the execution of the plan is starting. In other words, what conditions make up the raw material of this process you're about to design? Don't spend a lot of time on this step. Write a sentence or two, but not more than that. The rest will surface during the process of creating the tree. Examples might be:
   - In the case of creating a workshop that teaches transition trees, a starting point might be, "Students have never seen a transition tree before." An additional starting point might be, "Most of the students generally plan from a 'what to do' rather than a "'what to achieve' perspective.
   - In the case of the manufacturing company that was planning meetings for its personnel, a starting point was, "To avoid being blamed for other people's quality problems, operators normally ensure a batch is completed, and their machines cleaned up before the end of their shift." Another was, "Operators are concerned that if they really implement the new behaviors, they will lose a substantial amount of income that they now earn by working overtime."

d. **Decide upon an action that will lead toward achievement of the objective/s.** The question to answer now is, *What, when I do it, will get me closer to the objective?* I am using the term "I" here to be synonymous with the "creator of the tree." The creator of the tree may be an individual or a group of people. In any event, please note that every action in a transition tree is an action to be taken by the creator or creators of the tree. Any action that will be done by others should be an effect of an action taken by one or more of the tree's creators. The only exception to this is the case where others have previously given the tree's creators their permission to assign actions to them.

- When you actually sit down to do the transition tree, you may already have at least a partial plan in mind. You might already be thinking, *First I'll do this, then I'll do that, and then that, and then that.* If this is the case, take a moment to write down the actions that you already have in mind and put them aside. Without doing this, you will likely hold on to these ideas in your mind, distracting you from the process. Writing it down will free you to create the transition tree without clutter in your mind, and it will provide you with something to refer to for ideas as you move along in the process. Once you've written down what you already had in mind, select the first action that you plan to take.

Write the action in a square-cornered box, and verbalize it as a present tense action-entity. This means that instead of writing, "I will drive two miles north," you write, "I drive two miles north."

I'll sum up step one with a very simple example. Right now, I am sitting at my kitchen table, looking at a wall in my family room. The picture hanging on that wall is crooked, and I have just decided to straighten it. The wall is about 15 feet from where I sit. Let's say that for some very strange reason (like I must have *way* too much time on my hands), I decide to create a transition tree in order to plan how I'm going to get that picture straight.

Step 1a. This transition tree will describe how I'm going to go about getting that picture straightened.
Step 1b. My picture is hanging straight on the wall.
Step 1c. I'm sitting at my kitchen table, and that crooked picture is really annoying me.
Step 1d. I stand up.

2. **Using sufficient cause thinking, link the action to an objective.** Right now the only entities you have defined are an action and your objective/s. By following the steps below, you will determine additional actions and entities that will connect that initial action with all of the objectives. You are going to define the process that, when executed, will take you from where you are now to where you want to be.

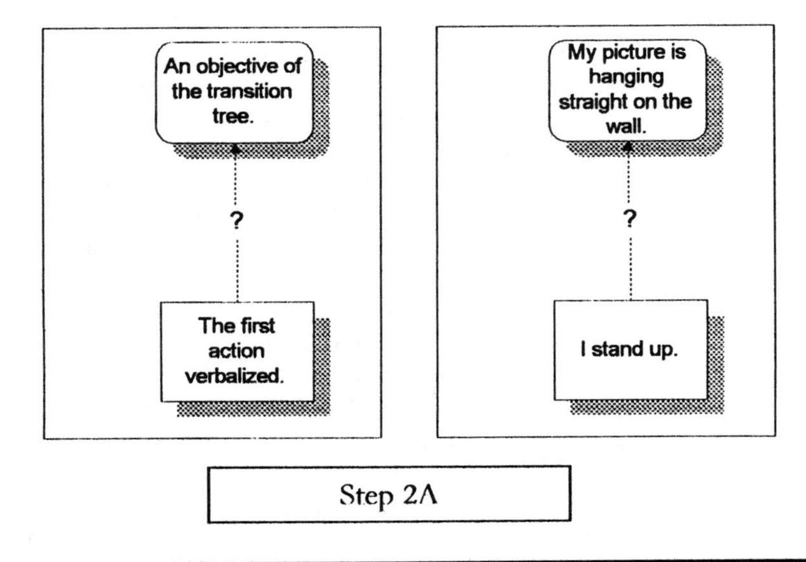

Figure 6.4

a. Try mentally drawing an arrow from the action directly to the objective (Figure 6.4). Ask yourself if the action is sufficient in and of itself to cause the objective. (*IF* [action], *THEN* [objective]). Well, if it were that simple, you should not have made the decision to use the transition tree.
b. Ask yourself the following question. *OK, so what does the action cause that is leading me toward the objective?* Referring once again to the manufacturing example, you are asking yourself to describe what the raw material has turned into as a result of the first step in the manufacturing process. Write that effect as an entity, and draw the arrow to it from the action. Your tree will now look something like Figure 6.5.
c. Subject this connection to the categories of legitimate reservation, and add any additional entities (including additional actions) that are necessary to clearly show sufficient cause as in Figure 6.6).
d. From this (or these) resulting entity (entities), repeat the same process.
   i. Try to connect what you have directly to the objective. If you prove to have sufficient cause (after utilizing the categories of legitimate reservation), great. If not,

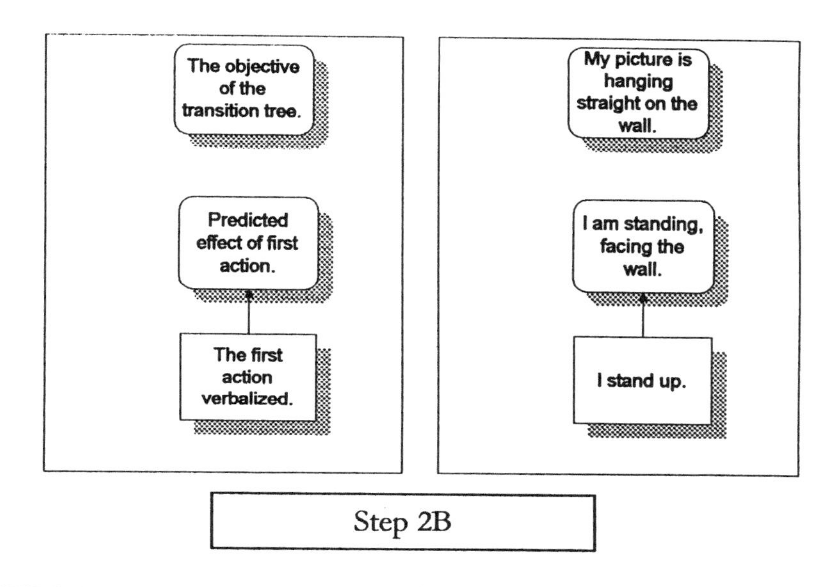

Figure 6.5

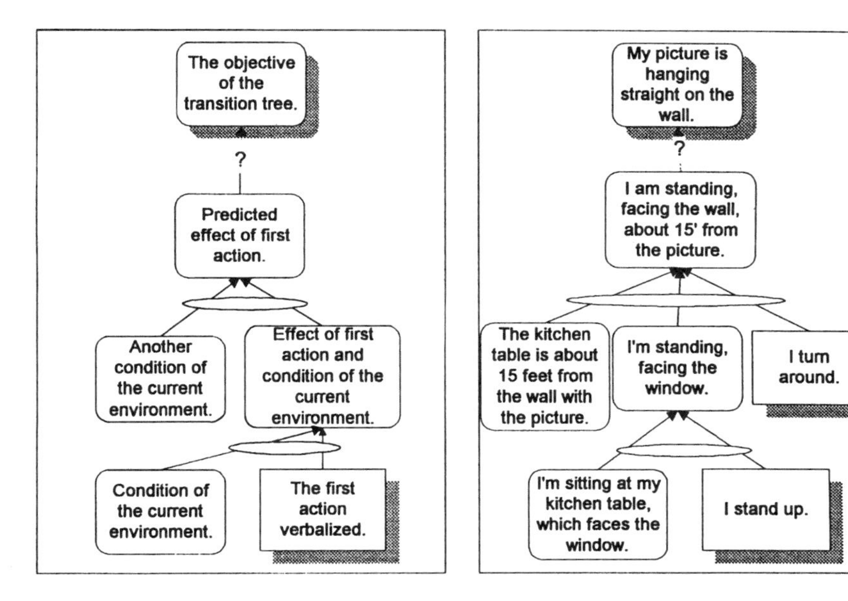

Figure 6.6

ii. Decide on the next action. What, when you do it, will create the next changes in the environment, leading toward achievement of the objective/s?
iii. Articulate the new effects of combining this new action, current conditions, and the conditions already defined in the tree cause.
iv. Return to step 2di, and repeat the process until all objectives are achieved.

My "straightening the picture" transition tree would now look like Figure 6.7.

Of course, in reality, I am not about to spend the time preparing a transition tree for a task as simple as walking across my family room to straighten a picture. However, I will suggest that when a transition tree is called for, so is the type of detail that you see in this simple example. To help explain why, let's return one more time to the manufacturing analogy. I spent several years in the cable assembly business, so I'll use that process as an example. The process for creating a coax cable assembly looks something like this:

1. Cut the cable to the specified length.
2. Strip ½" of jacketing material from each end.
3. Solder leads to individual wires.
4. Attach connector components.
5. Using potting compound, seal the interface between the connector and the cable.
6. Cure in oven for a period of time.
7. Perform electrical and mechanical inspections.
8. Prepare paperwork.
9. Product is now ready for customer.

It is easy enough to have the workers involved in the assembly process simply perform each step when the product comes their way. However, if they don't know what qualities the product is supposed to have when they start their step of the process, and when they end their step of the process, the chances of finishing with a high-quality product are diminished. If, on the other hand, the workers know what to look for each step of the way, they are much more likely to know when a problem exists and are better able to take corrective action — perhaps in enough time to make the customer's due date, and certainly before the customer finds the problem on his own. So, instead of simply stating the steps, manufacturers are learning to include a description of the state of the

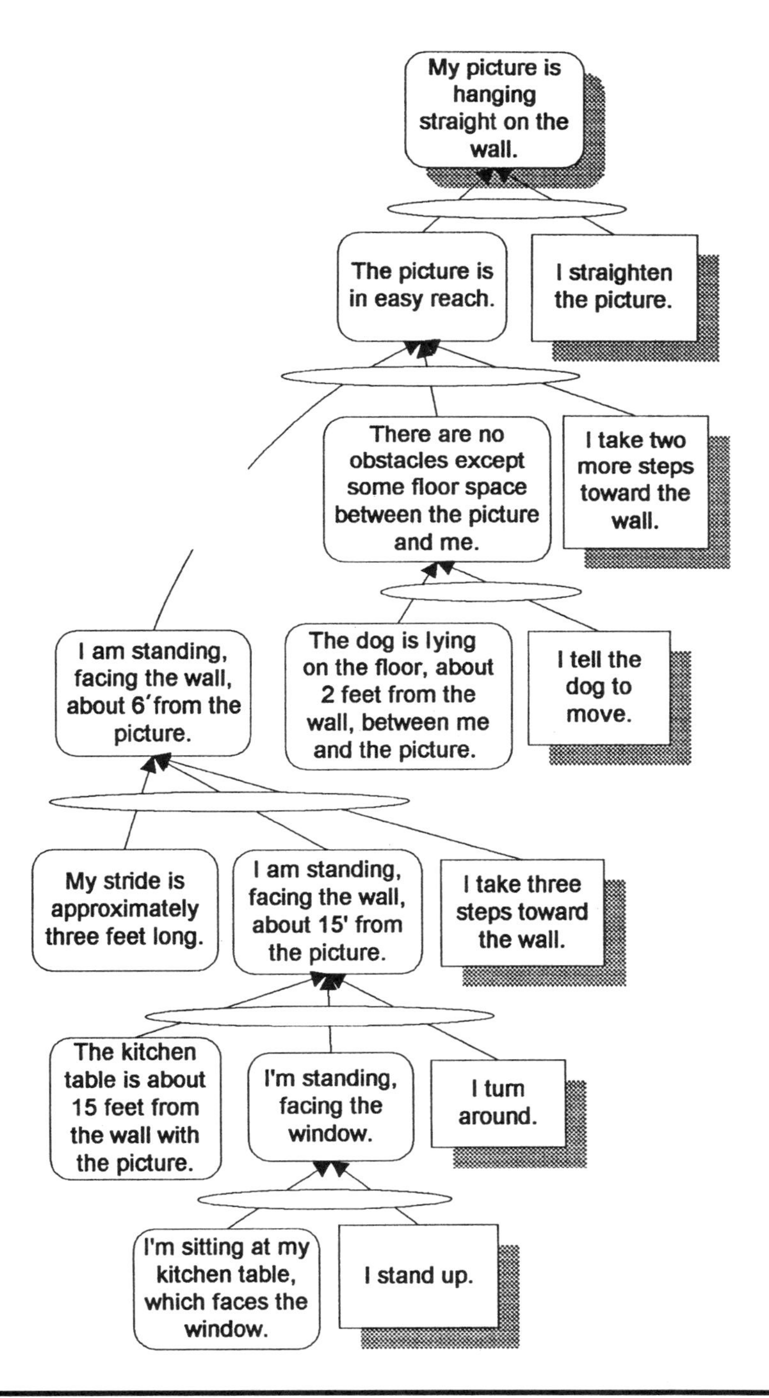
My picture is hanging straight on the wall.
The picture is in easy reach.
I straighten the picture.
There are no obstacles except some floor space between the picture and me.
I take two more steps toward the wall.
The dog is lying on the floor, about 2 feet from the wall, between me and the picture.
I tell the dog to move.
I am standing, facing the wall, about 6′ from the picture.
My stride is approximately three feet long.
I am standing, facing the wall, about 15' from the picture.
I take three steps toward the wall.
The kitchen table is about 15 feet from the wall with the picture.
I'm standing, facing the window.
I turn around.
I'm sitting at my kitchen table, which faces the window.
I stand up.

Figure 6.7

product at the beginning and end of each step. The description of the first two steps of our coax cable assembly process might then look like:

1. Cut the cable to the length specified. Cable should be at length, ±.05".
2. Strip ½" of jacketing material from each end. Each end of the cable should now have ½", ±.025" of wires exposed. Individual wires should have no nicks from the stripping tool.

Also, in manufacturing, you know you can't take step three until the results of step 2 are accomplished. This is because the material does not yet meet the conditions required for that next step in the process to create the conditions that it is meant to create. The same holds true for the plans we make. In the simple transition tree, for instance, if my stride was only two feet, after three steps I'd be 9 feet from the wall instead of 6. Based on that, I would need to make course correction (like take a couple extra steps) in order to reach the wall and be able to straighten the picture. If the dog decided to get up and move on his own, I wouldn't need to tell him to move — another course correction, based on the existence (or non-existence) of the conditions that were presumed to exist in the current reality, or the conditions that were predicted.

Unless we train ourselves to look for those conditions, we will continue to waste time doing, and not achieving.

- **Helpful Hint:** When you are predicting an intangible effect, utilize the predicted effect reservation to identify observable effects of the intangible. How do you know that an entity that you cannot observe exists? You know only by the observable, predicted effects of that entity. For instance, let's say that your action is, "I tell a joke," and the effect of telling the joke is that "my audience is warming up." How do you *know* that the audience is, in fact, warming up? You look for *its* effects. They're smiling, or they're laughing. By putting these predicted effects on your tree, you will be able to ensure that they exist before moving on to the next action. The importance of this hint will become even clearer in step four.

3. **Seek and block undesirable consequences.** At this point, you have a transition tree that describes a plan for how you are going to successfully reach Point B from Point A. In this step, you are going to review your plan (transition tree) and look for potential side effects. Assuming that everything happens as described in the transition tree, will anything *else* happen, too? In particular, will anything else happen that you don't want to happen?

In each action we must look beyond the action at our past, present, and future state, and at others whom it affects, and see the relations of all those things. And then we shall be very cautious.

Pascal, 1670

---

a. Identify and list any potential undesirable effects of any of the injections, entities, and causal relationships developed in the tree. Consider potential impact on stakeholders: owners, employees, vendors, customers, community, family, and friends.
   i. If the plan that is described by the transition tree is one that will involve significant risk, resources, or time, I strongly urge you to also ask a colleague to review it in order to identify such undesirable effects.
   ii. If none is identified, continue with step four. Otherwise, move on to the next step, 3b.
b. Select one of the potential undesirable effects and identify the entity or entities that will cause it to exist. Add the entity (the undesirable effect) to the tree. Your diagram will resemble Figure 6.8.
c. Scrutinize the relationship using the categories of legitimate reservation, and modify as needed to solidify the connection. Your tree may now look something like Figure 6.9.
d. Determine an action (injection) that will block the emergence of the undesired consequence. What, if you did it, would prevent the undesirable effect from emerging? Remember the simpler the better. You are still trying to create the shortest, albeit pain-free, path from Point A to Point B.
e. Inserting the new injection, recreate this portion of the tree. The undesirable effects that are blocked by this injection should not be on this diagram. Instead, you should see desirable (or at least non-undesirable) entities in their place (Figure 6.10).
f. It is quite possible that the decisions you have just made, as reflected in the modifications you have made to your tree, are sufficient to deflect one or more of the remaining predicted consequences. If so, by all means strike them from your list!
g. Repeat steps 3a through 3f until you have resolved all of the undesirable consequences that you are able to predict.

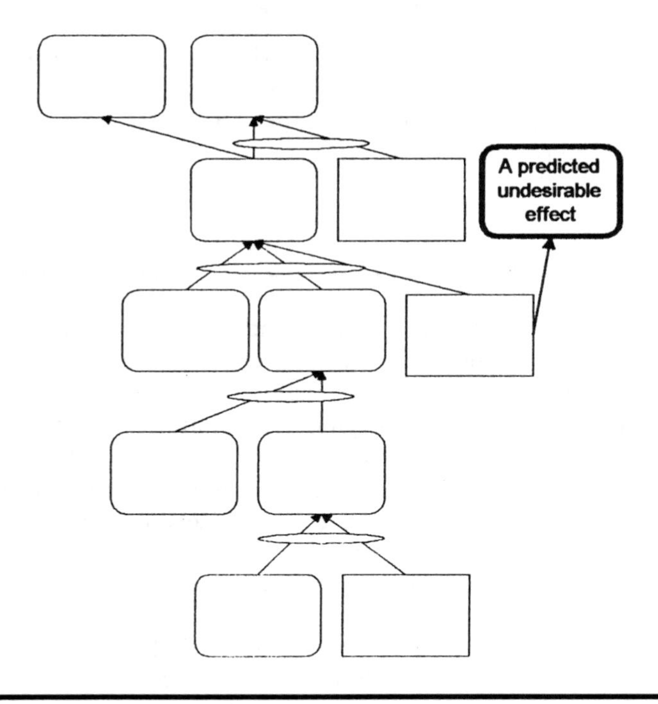

Figure 6.8

4. **Implement the plan!** Chances are, you won't get from Point A to Point B unless you start to take action! Implement the first action defined on your transition tree. When the effects of that initial action are achieved, it's time to take the next action. When the effects of that action are achieved, move on to the third action, and so on. What if you have taken action #2, and the predicted effects don't happen?
   - First, make sure you've given the effects enough time to occur. If you believe that you have waited long enough, then it's time to revise the plan. Yes, in the real world, we can't predict everything with great accuracy. Sometimes, reality changes, and we must change our carefully laid plans along with it. The first option is to take another action or set of actions to create the effects that are required before you move forward with action #3. The second is to redo the transition tree from your new starting point (Figure 6.11).

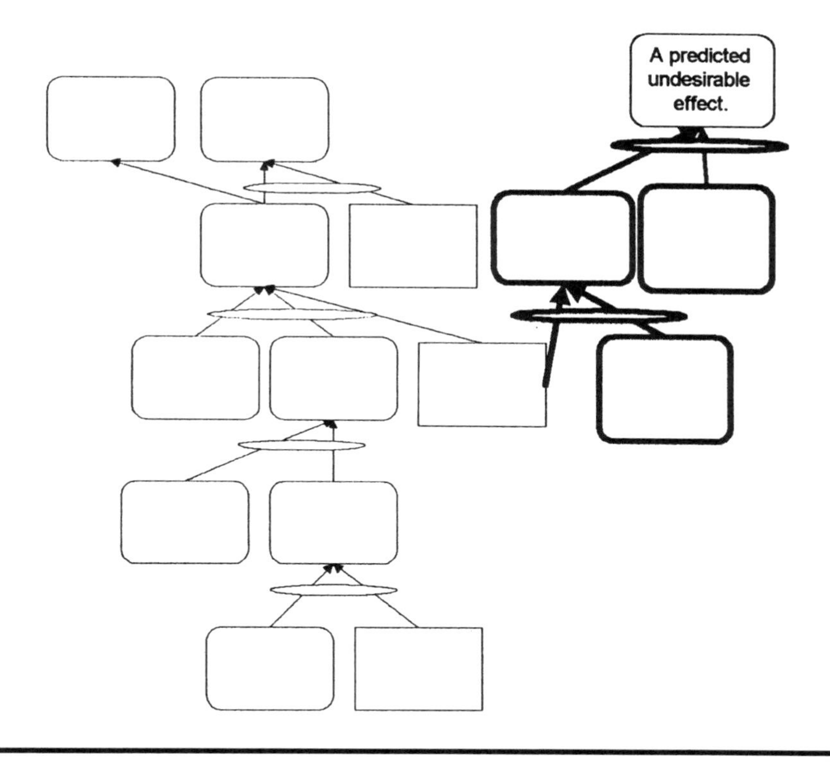

Figure 6.9

An Example

The following pages contain an example of a transition tree that I created to prepare for a presentation I was asked to give at a conference. This particular conference is held annually and is devoted to the subject of "cost." The vast majority of the sessions at this conference provide attendees with information on activity-based costing (ABC), and activity-based management (ABM), which represent the current popular methods for tracking and allocating costs. From a TOC perspective, the whole notion of allocating costs that are not totally variable to the unit of sale is considered at best a waste of time, and at worst a major obstacle to the continued financial health of organizations.*

* The entire discussion of this subject is outside the scope of this book. If you're interested in reading more about the TOC perspective on management accounting, start with Dr. Eliyahu Goldratt's *The Haystack Syndrome* (North River Press, 1990) and *The Theory of Constraints and Its Implications for Management Accounting* (North River Press, 1995) by Eric Noreen, Debra Smith, and James T. Mackey.

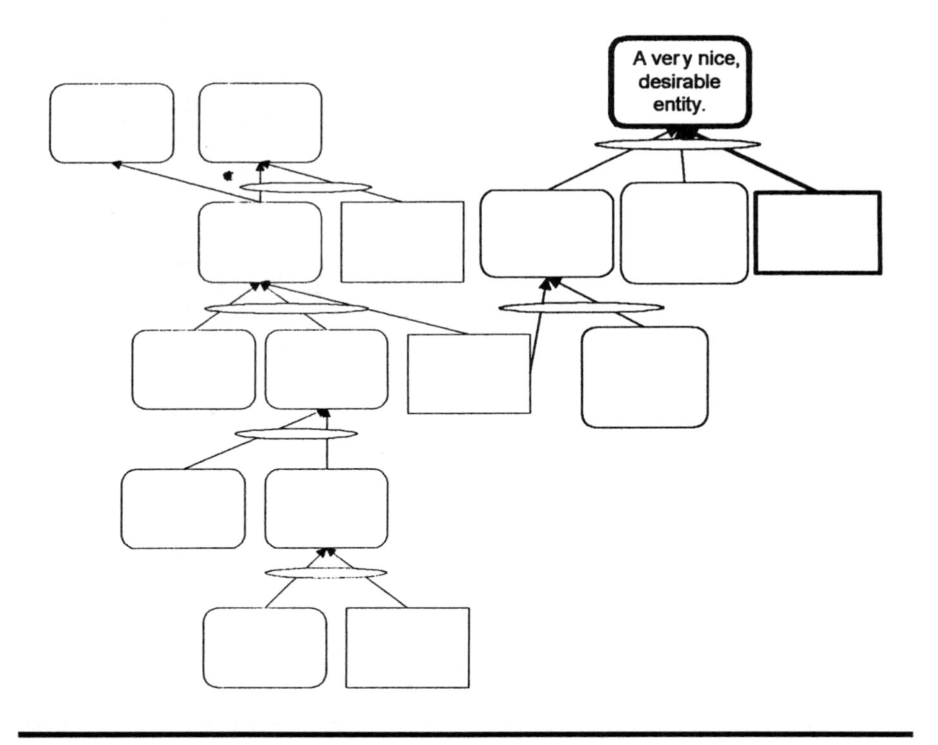

Figure 6.10

I was asked to make a presentation on the TOC perspective, as "The Contrarian." In addition, I was asked to participate in a panel discussion (also as the lone "contrarian") on the topic as well. My presentation was to be one hour long, including questions and answers. I decided to use a transition tree to plan my presentation.

My first task was to define an objective. After careful thought, I decided that in one hour with a likely hostile audience, an achievable objective could be, "They want to learn more about TOC." I knew that in order to accomplish that, I would need to do a few things. I would need to establish some credibility with the audience. They would need to know that I had at least some understanding of the good things they are trying to do for their organizations and the hard work it takes to do what they do. The transition tree shows how I accomplished that. The transition tree that you see didn't start out looking like the final product. I went through several iterations, and made sure to get feedback and help from a colleague, Chris Waddell, who is an accountant. The evaporating cloud that I used in the presentation also started out worded quite differently.

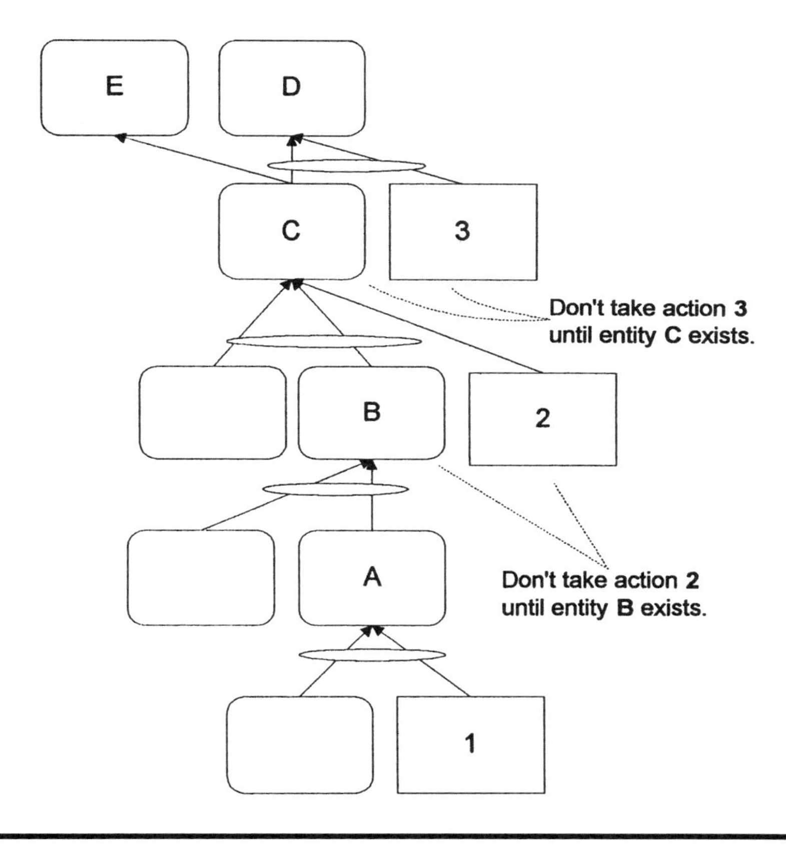

Figure 6.11

Again, Chris provided the insights and perspective of the people who would be my audience.

I did, in fact, achieve the objective of the presentation. Many of the people who attended the session stayed over to ask questions. Many of them have since gone on to study TOC, some with me, and some through other channels. As a result of the presentation, I was asked to facilitate a few meetings of the local IMA (Institute of Management Accountants) breakfast club, which was a special interest group studying activity-based costing.

I also used the transition tree as a learning tool for myself. After the session, I reviewed the tree and identified where I thought the presentation could have been improved. Either by cutting injections because the effects they were meant to cause had already existed, or by adding clarity to others, to ensure smoother transitions. This process has helped me do a better job when speaking, because I use each event and the transition tree associated with it as a personal learning opportunity.

Please note the direction of the arrows when you read this tree (Figure 6.12). The bottom of the tree is at the top of the page. Get in the habit of looking for the direction of the arrows rather than where entities are located physically on a page or computer screen.

As you read through the tree, practice your use of the categories of legitimate reservation. Wherever you have a question with an entity or relationship, make the changes that are necessary for the relationships to be valid. What assumptions must I have made, that perhaps aren't articulated? What assumptions are you making when you find yourself in disagreement? What must have existed, that may not be spelled out clearly on the tree, that enabled the presentation to be a success?

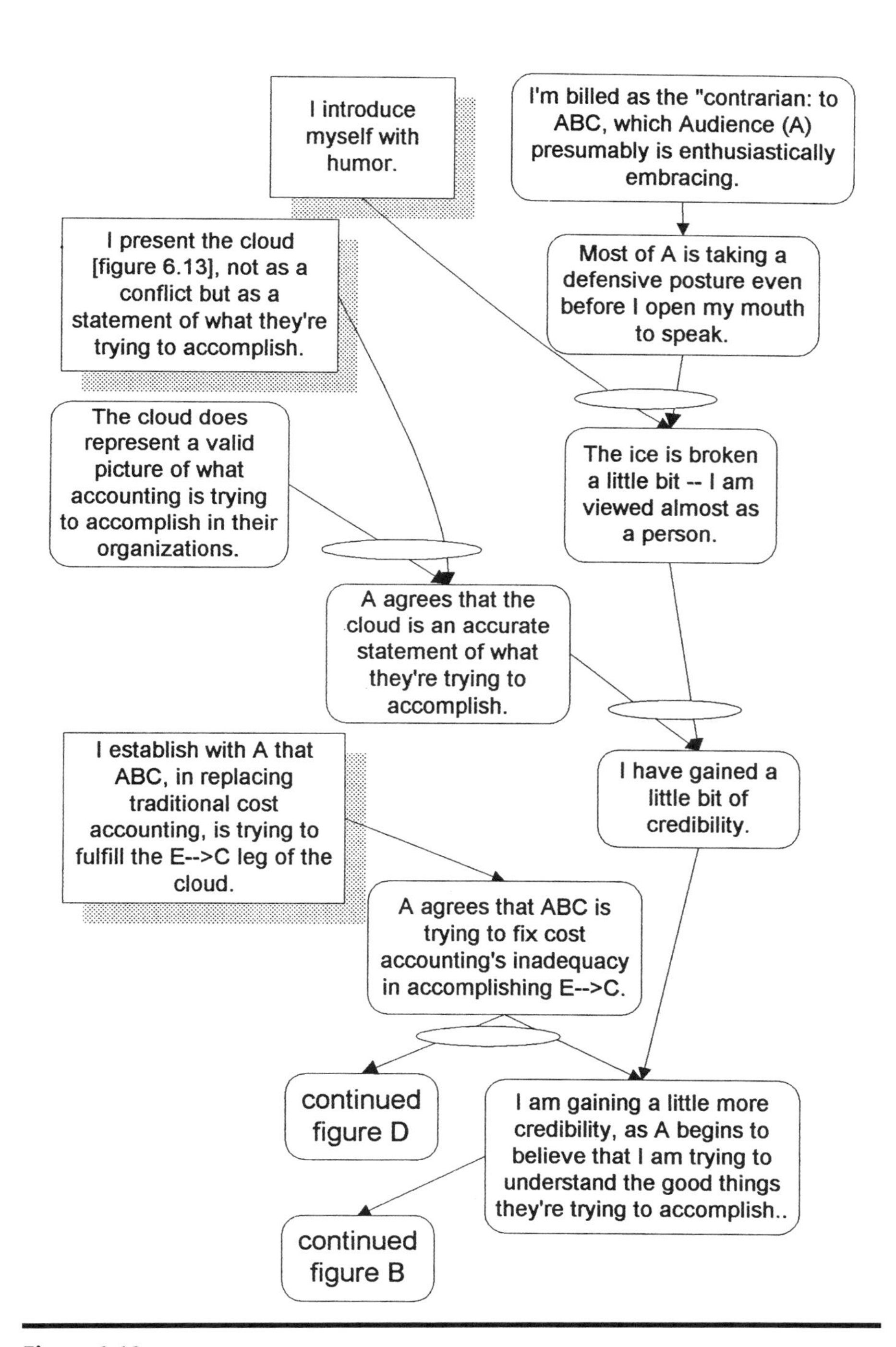
I introduce myself with humor.
I'm billed as the "contrarian: to ABC, which Audience (A) presumably is enthusiastically embracing.
I present the cloud [figure 6.13], not as a conflict but as a statement of what they're trying to accomplish.
Most of A is taking a defensive posture even before I open my mouth to speak.
The cloud does represent a valid picture of what accounting is trying to accomplish in their organizations.
The ice is broken a little bit -- I am viewed almost as a person.
A agrees that the cloud is an accurate statement of what they're trying to accomplish.
I establish with A that ABC, in replacing traditional cost accounting, is trying to fulfill the E-->C leg of the cloud.
I have gained a little bit of credibility.
A agrees that ABC is trying to fix cost accounting's inadequacy in accomplishing E-->C.
continued figure D
I am gaining a little more credibility, as A begins to believe that I am trying to understand the good things they're trying to accomplish..
continued figure B

Figure 6.12a

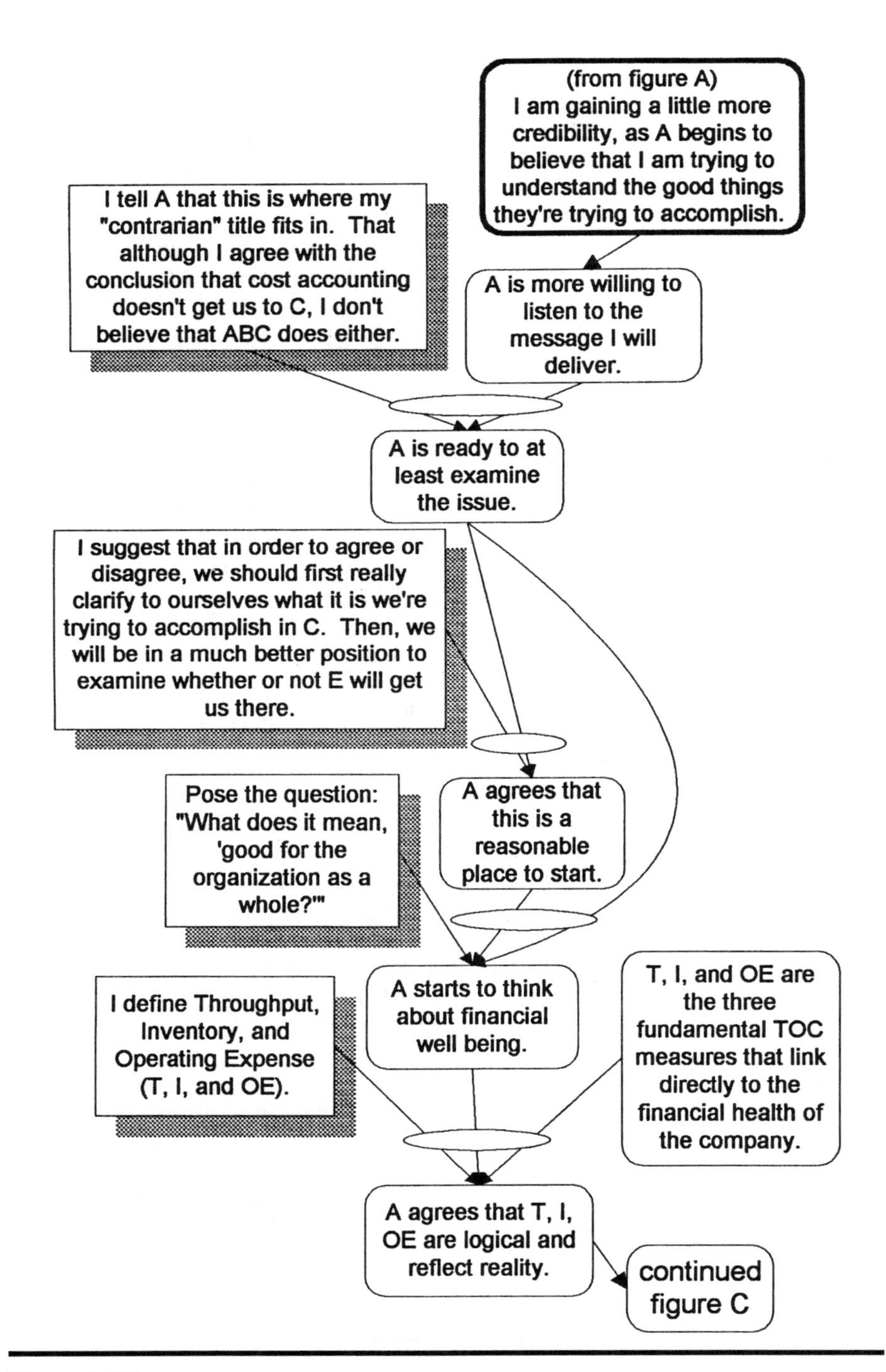
(from figure A)
I am gaining a little more credibility, as A begins to believe that I am trying to understand the good things they're trying to accomplish.
I tell A that this is where my "contrarian" title fits in. That although I agree with the conclusion that cost accounting doesn't get us to C, I don't believe that ABC does either.
A is more willing to listen to the message I will deliver.
A is ready to at least examine the issue.
I suggest that in order to agree or disagree, we should first really clarify to ourselves what it is we're trying to accomplish in C. Then, we will be in a much better position to examine whether or not E will get us there.
A agrees that this is a reasonable place to start.
Pose the question: "What does it mean, 'good for the organization as a whole?'"
A starts to think about financial well being.
I define Throughput, Inventory, and Operating Expense (T, I, and OE).
T, I, and OE are the three fundamental TOC measures that link directly to the financial health of the company.
A agrees that T, I, OE are logical and reflect reality.
continued figure C

Figure 6.12b

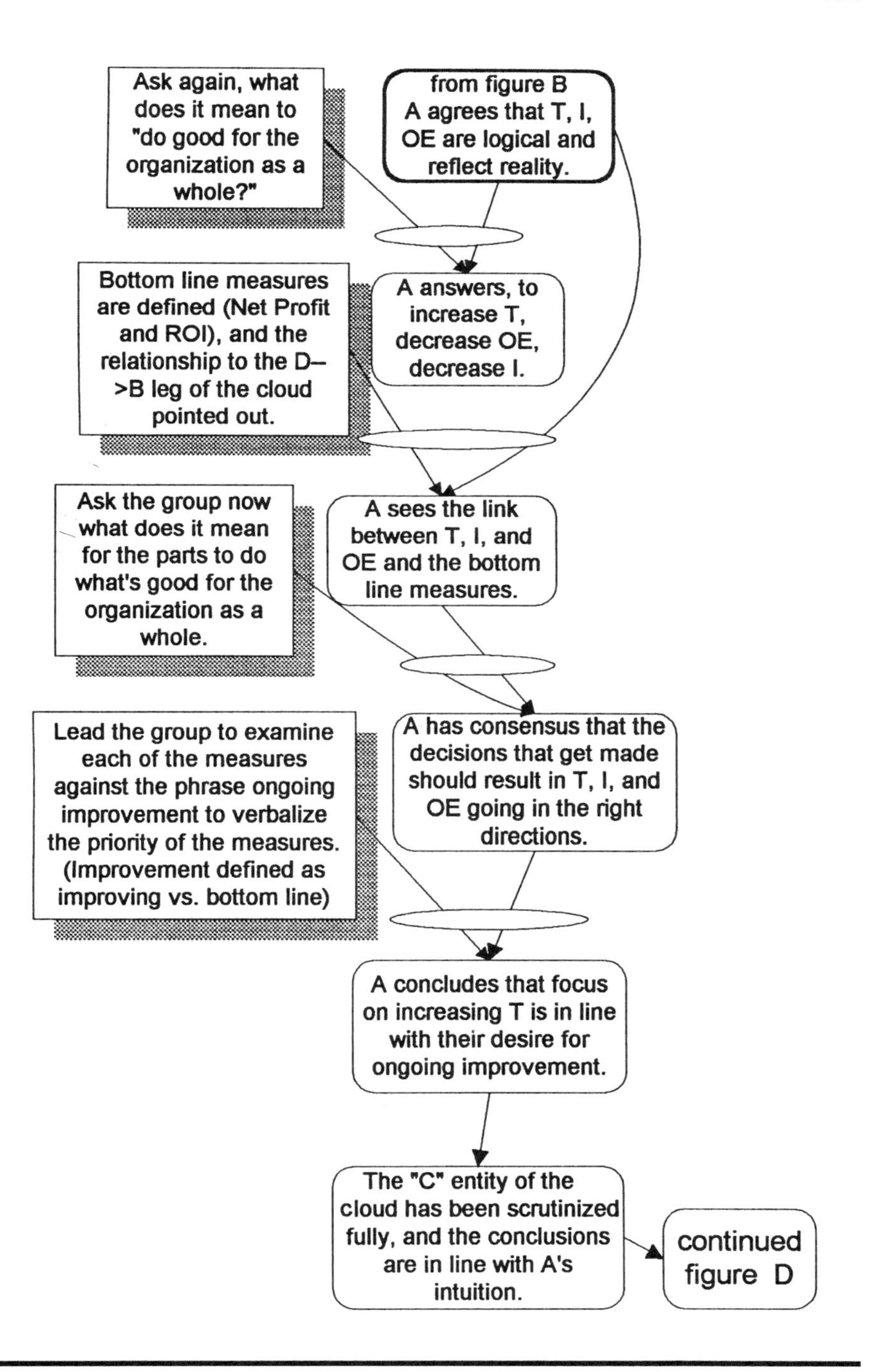

Ask again, what does it mean to "do good for the organization as a whole?"
from figure B A agrees that T, I, OE are logical and reflect reality.
Bottom line measures are defined (Net Profit and ROI), and the relationship to the D-->B leg of the cloud pointed out.
A answers, to increase T, decrease OE, decrease I.
Ask the group now what does it mean for the parts to do what's good for the organization as a whole.
A sees the link between T, I, and OE and the bottom line measures.
Lead the group to examine each of the measures against the phrase ongoing improvement to verbalize the priority of the measures. (Improvement defined as improving vs. bottom line)
A has consensus that the decisions that get made should result in T, I, and OE going in the right directions.
A concludes that focus on increasing T is in line with their desire for ongoing improvement.
The "C" entity of the cloud has been scrutinized fully, and the conclusions are in line with A's intuition.
continued figure D

Figure 6.12c

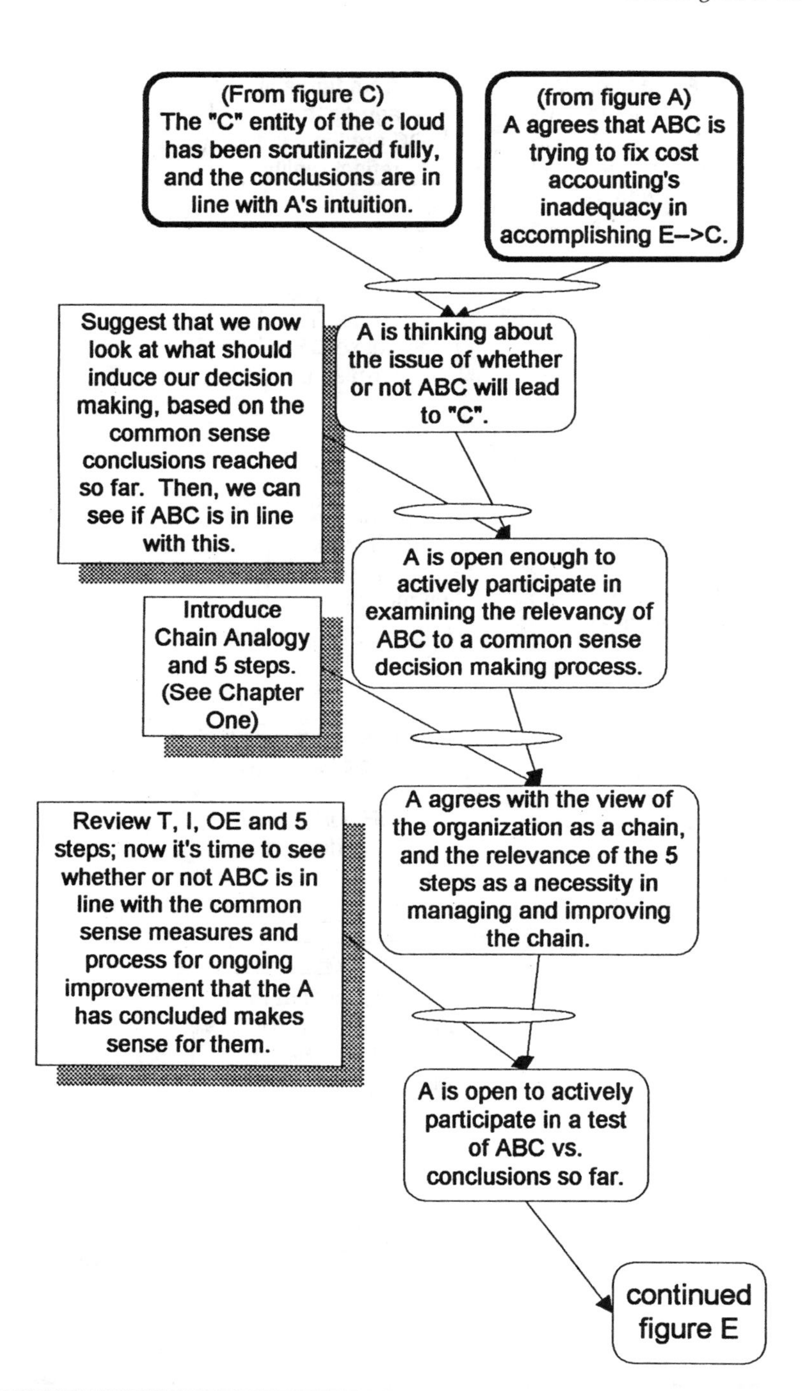
(From figure C)
The "C" entity of the c loud has been scrutinized fully, and the conclusions are in line with A's intuition.
(from figure A)
A agrees that ABC is trying to fix cost accounting's inadequacy in accomplishing E–>C.
Suggest that we now look at what should induce our decision making, based on the common sense conclusions reached so far. Then, we can see if ABC is in line with this.
A is thinking about the issue of whether or not ABC will lead to "C".
Introduce Chain Analogy and 5 steps. (See Chapter One)
A is open enough to actively participate in examining the relevancy of ABC to a common sense decision making process.
Review T, I, OE and 5 steps; now it's time to see whether or not ABC is in line with the common sense measures and process for ongoing improvement that the A has concluded makes sense for them.
A agrees with the view of the organization as a chain, and the relevance of the 5 steps as a necessity in managing and improving the chain.
A is open to actively participate in a test of ABC vs. conclusions so far.
continued figure E

Figure 6.12d

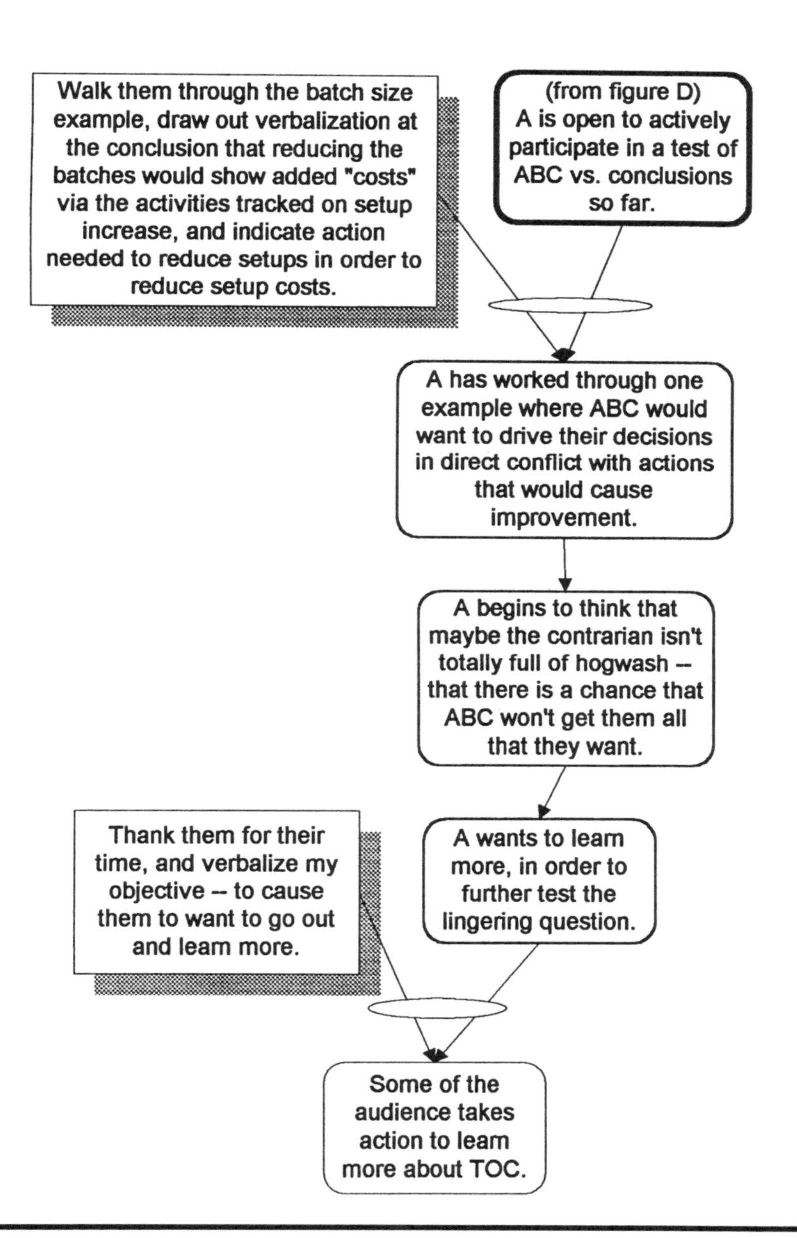
Walk them through the batch size example, draw out verbalization at the conclusion that reducing the batches would show added "costs" via the activities tracked on setup increase, and indicate action needed to reduce setups in order to reduce setup costs.
(from figure D) A is open to actively participate in a test of ABC vs. conclusions so far.
A has worked through one example where ABC would want to drive their decisions in direct conflict with actions that would cause improvement.
A begins to think that maybe the contrarian isn't totally full of hogwash -- that there is a chance that ABC won't get them all that they want.
Thank them for their time, and verbalize my objective -- to cause them to want to go out and learn more.
A wants to learn more, in order to further test the lingering question.
Some of the audience takes action to learn more about TOC.

Figure 6.12e

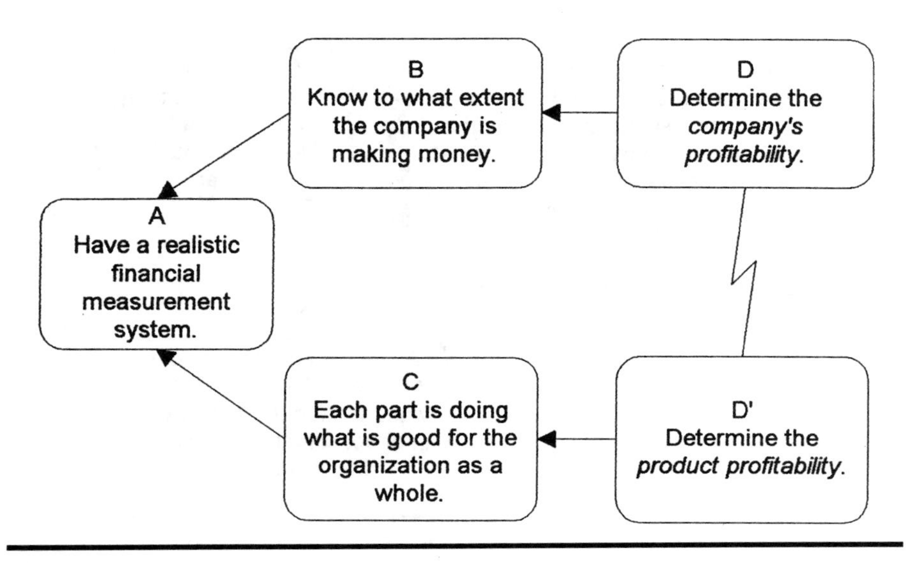
B
Know to what extent the company is making money.
D
Determine the *company's profitability*.
A
Have a realistic financial measurement system.
C
Each part is doing what is good for the organization as a whole.
D'
Determine the *product profitability*.

Figure 6.13

Chapter 7

Future Reality Tree

> You choose, you live the consequences. Every yes, every no, maybe, creates the school you call your personal experience.
>
> Richard Bach, 1994

Every single decision we make, every single action we take, will change something in the future. It doesn't matter whether that future is just a moment away or whether that future is years from now. The change might be small enough that it's barely noticeable, or so large that an entire civilization is affected.

As its name suggests, the future reality tree is a tool for visualizing and predicting the future. But, you may say, the future is unpredictable! True. A butterfly flapping its wings in China may cause entire weather patterns to change in North America six months from now.

Yet don't we attempt to predict the future with *some* degree of accuracy every day? You set your alarm clock before you go to bed at night, predicting that it will ring at the appointed time and predicting that as a result, you will wake up in time to do whatever it is you will do that day. You decide to hire a particular job candidate because you predict that she will fit into the company culture and perform the tasks of the job with a high degree of proficiency. You decide *not* to hire a particular job candidate because you believe that bringing that person up to speed on your technology will take an inordinate amount of time. Your company develops a new marketing strategy, predicting that as the strategy is implemented, market share will increase, along with profitability. You bring a gift home for your spouse, predicting that he or she will be happy as a result. You put your young child in "time out" for 10 minutes, predicting that as a result, she'll stop the inappropriate behavior. Professors predict that students will learn as a result of implementing their lesson plans. Sales professionals predict they will "close" business when they deploy their sales skills and plans. A manager is afraid to enforce a required new behavior in his plant, because he fears the workers will retaliate. A child is afraid to raise her hand in class because she thinks she'll look stupid and her friends will make fun of her.

Yes, we are predicting the future all the time. Sometimes, we do so quite inaccurately. Always we do so quite incompletely. All too often, if we'd have just *thought* about it a little bit, we could have predicted better — more accurately and more completely. The future reality tree is a tool that helps us do so, and has the important added benefit of helping us learn more about our changing systems as we go.

Future reality trees are sufficient cause diagrams that contain four distinctive parts, which are labeled in Figure 7.1.

A. **Injections** are always entry points to the tree (see Chapter 4). Injections are entities that do not exist in the system's current reality, and are distinguished from other entities by their squared corners. Why the term *injection*? Think of getting a shot in the arm. The idea is that once you've received that injection, the illness will be cured, and the ugly symptoms will disappear. Once an injection (idea) is implemented in reality, the effects predicted should emerge as a result.
B. Entities that do currently exist in the system's reality. In a future reality tree, this type of entity will usually be entry points and is typically not found in the body of the tree.
C. Entities that do not yet exist in the system. When entities that currently exist (B) are combined with injections (A), the (C) entities will (at least they're predicted to) exist in the future.
D. Reinforcing loops are often placed in future reality trees, as a means to create patterns of sustained and continuous improvement.
   - The key to creating the desired future reality is implementing the injections.

When to Use the Future Reality Tree

Any time it's important to put some thought into the web of effect–cause–effect connections among one or more ideas and the predictable

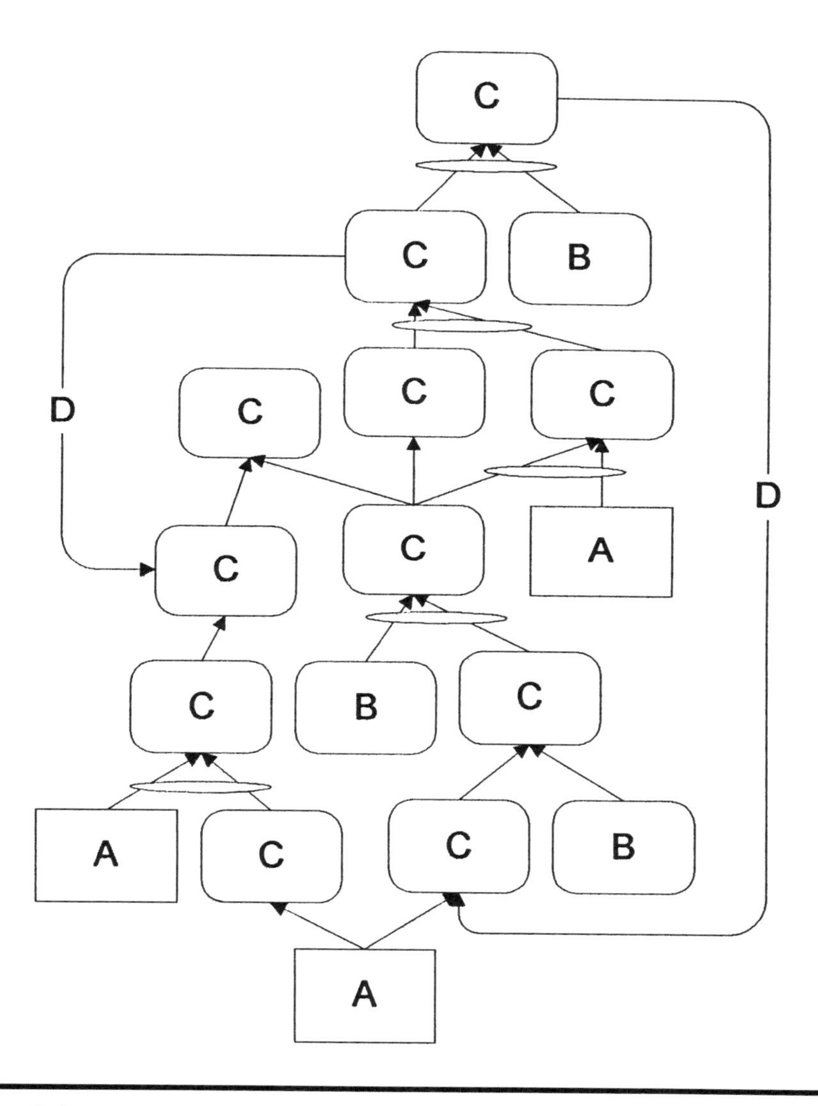

Figure 7.1

inevitable consequences (effects) of those ideas, it's time to reach for the future reality tree tool. Some general applications of the future reality tree include:

- When you haven't yet decided what to implement, in order to create a change.
- When you want to explore the potential effects of an idea, before implementing it.

- When you want to see if, in fact, your idea can actually achieve what it's meant to achieve.
- When you want to see what else to implement, in addition to an initial idea, in order for all objectives to be achieved.
- When you think that an idea might have undesirable effects (unintended, problematic consequences).

More specifically, future reality trees are useful when applied in situations such as:

- When a policy decision is going to be made in one department, it might be a good idea to try to understand what the effects of that decision will be on that department, other departments, customers, employees, suppliers, stockholders, and other stakeholders.
- When that which you want to implement has the likelihood of requiring the organization's resources, such as money, people, time, or technology.
- When a government wants to add, change, or remove a law, and the possibility exists that the law could have effects on other aspects of the community.
- When a team wants to put together a cohesive, clear vision of the future.
- When a colleague has an idea that you think is a lousy one, and you want to communicate your concerns in a constructive, non-threatening way.

The Process

The future reality tree process consists of three main steps.

1. Define the basis for the tree.
   a. Identify an injection (idea).
   b. List the objectives (pro's) of the injection.
   c. List potential undesirable consequences (con's) of the injection.
2. Describe the effect–cause–effect relationships.
   a. Using sufficient cause thinking, connect the injection to the objectives.
   b. Using sufficient cause thinking, seek and block potential undesirable consequences of the injection.
3. Enhance the solution.
   a. Predict additional effects.
   b. Add reinforcing loops.

After I go through each of these steps in detail, I will provide some examples of real people's real future reality trees.

Do you have any ideas that you're contemplating right now? Pick one, and do the tree while we go through the steps that follow.

1. **Define the basis for the tree.** Every future reality tree begins with the following three elements. The three elements combined answer the question, *So what's the big idea, and what do I/we expect it will accomplish?* The order in which you establish them depends on the situation. It's OK for any of the three to be your starting point, but don't move to step two until you've defined all three elements that comprise the basis for the tree.

   I often find it helpful to give myself the opportunity to "tell the story" before I worry about specifically articulating the entities that comprise the basis for the tree. Once I've scratched out a paragraph or two that describes what I'm expecting, I then move on to the discipline prescribed by the process. This allows me the freedom to derive the benefits from both free-form thinking *and* technical correctness.

   a. **Identify an initial injection (idea).** What is it that you are going to consider implementing? Remember that the injection is an entity and should be written as a complete sentence. Writing the injection in the present tense helps you mentally project yourself into that future state. As you progress with the tree, using the present tense makes the process of testing assumptions (with the categories of legitimate reservation) easier. Initial injections may come from several potential sources, depending on what led you to creating the tree in the first place:

      - The evaporating cloud process
      - A brainstorming session
      - An idea that a colleague has presented to you
      - An idea that just comes into your head (have you ever had a fantastic idea in the middle of the night, or while in the shower?)
      - A suggestion from your boss
      - A suggestion from a family member
      - Thin air

      Here are some examples, along with the ways in which they'd be verbalized and drawn as initial injections. All three are examples of injections — ideas that *somebody* wanted to make happen. Most often, the *idea* — the *injection* (whether it's yours, a colleague's, or someone else's idea that has culminated in an opinion from you) — is the cognitive starting point of the future reality tree.

- In an attempt to stun the competition, capture a larger market share, and increase its profits, a firm that manufactures printed circuit boards decides to offer a three-day lead time for prototype jobs.
- A candidate in the 1996 presidential election campaign suggests that an across-the-board, flat tax should replace the current income tax structure in the United States.
- You think it would be a great idea to go on a Caribbean cruise with your spouse. You both have been working hard, and just the thought of sun, fun, and ocean are already making you feel good.

We are offering a three-day lead time for prototype jobs.

The U.S. Has an across the board, flat tax.

My spouse and I are on a Caribbean cruise.

Figure 7.2

b. **List the objectives (pro's) of the injection.** People come up with new ideas and then believe in them because they trust that good things will happen once their ideas turn into reality. What are the predicted positive effects of the injection? What does the person (or group) who has the idea believe will happen as a result of having the idea implemented? Even if *you* believe your colleague's idea is a lousy one, and all you can think of initially are the reasons the idea shouldn't be implemented, it is important that you take a few moments to contemplate why *they* believe it's a good idea. Remember — it's all in the assumptions, and you are making them, too! **State each objective as an entity and verbalize in present tense terms**.

- The printed circuit board company was already a valued source of supply for high-volume orders. However, they knew that their customers also often went through a somewhat painful and costly process when launching new designs and products. There are many small shops that specialize in turning around prototypes of new product designs quickly, and they command high prices for doing so. This manufacturer believed that by offering a three-day lead time for these higher-priced prototype jobs, it would increase its profitability. It also believed it would increase customer loyalty and capture a larger share of the market, because it would be offering one-stop shopping to those clients, who would choose to keep their business with the company that worked with them to perfect the design. The management team may have verbalized the objectives as follows:

- Customers are gladly paying X% over "standard price" for three-day turnaround products.
- We are enjoying significantly higher profits.
- We own a larger share of the market.
- They may verbalize the above more definitively by stating something like, "We own 80% of the market for prototype work in northern California."

- The presidential candidate suggests that the across-the-board, flat tax he is promoting would be more fair than the current structure, would generate more income for the country, need much less administration on the part of the IRS, and thus save money by vastly downsizing the IRS. If he had prepared a future reality tree, he may have stated the objectives of his idea as follows:
  - The country enjoys a fair tax structure.
  - The flat tax system is easy to administer.
  - The IRS has X number of employees.
  - The country is spending $X on administering its income tax system.
- Can you think of the reasons why a couple might want to go on a Caribbean cruise? List those reasons here, and then turn them into entity statements.

______________________	______________________
______________________	______________________
______________________	______________________

Another situation that falls into this category is when you follow a current reality tree with a future reality tree. (See Chapter 8 for a complete discussion of the Current Reality Tree.) When you have used the current reality tree to identify a core problem, you have created a model of the interrelated, undesirable aspects of a system. The current reality tree is describing the system (or at least the part of the system) that you would rather live without. The future reality tree is used to describe the system that you *do* want. So, you must verbalize the entities that will, in the future reality, replace the undesirable entities of the current reality. One way to go about defining these entities is to ask, for each of the pertinent (reselected) undesirable entities, *What is its opposite?. What would I rather see in its place?*

For instance, let's say that you are a university professor and you have prepared a current reality tree to determine the core problem of your classes. One of the reselected undesirable entities is, *On average, my classes suffer from a 40% dropout rate within the first month*. You would ask yourself specifically, what would you rather be experiencing in the future? What would you consider to be its opposite? You might decide that the entity of the future will be, *My classes enjoy a 0% dropout rate*. You might decide to verbalize the entity of the future as, *On average, my classes have a 5% or lower dropout rate*. Or you might decide that it should be, *The dropout rate of my classes is steadily improving*. The right answer is the entity you decide you want to create in your future.

A list of objectives may be the starting point of a future reality, rather than an injection. You may know the good things you'd like to see in the future, but haven't yet turned that list of good things into any sort of interconnected vision. As I write, I am in the first step of defining, with one of our clients, the type of relationship our firms will have in the future. In order to create our vision and then reach agreement on what the relationship will actually be, we are beginning with a future reality tree. Both parties are submitting five objectives that they want to see as a result of the relationship.

Do not compromise your objectives! People often lower their sights because they assume, before they even allow themselves to explore a little further, that they won't be able to achieve what they'd really like to achieve. Don't let your "can't do" preconceptions block you from envisioning great things. Use the future reality tree to create the vision to which you aspire. Address the obstacles with the prerequisite tree, and you'll see it's much easier than you think to reach for and actually capture the stars!

c. **List potential undesirable consequences (con's) of the injection.** Now comes the part of the process in which most of us really excel, especially *when the injection is someone else's idea*. Our task here is to identify everything that might go wrong as a result of implementing the injection. The famous, "Yes, but..." I know you've heard those two dreaded words, and it's likely that you have uttered them more than once yourself. They can be disguised as other phrases of objection, such as

- What a lousy idea!
- No way. Absolutely no way!
- Are you nuts?!?!?
- Hmmnn, let me think about it…
- Do you really think that's such a good idea…?
- I'm not sure about this.
- That'll never happen here!
- I'm afraid the cure might be worse than the disease.

There are two types of *Yes, buts*. One occurs when you are concerned that the injection itself might be a cause for undesirable effects. *"If that injection actually becomes reality in the future, bad things will happen as a result."* This is the type of "yes, but" you should be concerned with when creating a future reality tree. It is called a *negative branch reservation*. The other type of "yes, but" is an *obstacle*. Obstacles are objections that describe reasons why it will be difficult to make the injection a part of reality. When these types of objections surface, jot them down and put them aside until after you have completed the future reality tree. Then, use the prerequisite tree to determine how to remove them as obstacles to your desired future. (See Chapter 10 for complete instructions on the Prerequisite Tree process.)

It is certainly possible for the same "yes, but" to be a negative branch or an obstacle. Do you remember the Caribbean cruise idea I mentioned earlier? "Yes, but, it's very expensive!" Am I talking about a negative branch or an obstacle here? Depending on what the rest of my thinking is behind the statement, it can be either. By adding some clarity, I can easily determine whether I'm concerned with an obstacle or a negative branch.

Obstacle: It's too expensive! There's no way we can go on a cruise, because we don't have enough money to pay for the trip!

Negative Branch: It's too expensive! If we spend that kind of money, we'll end up in debt! Where will we find the money for next month's mortgage?

Identifying "yes, buts" is crucial. Would you rather find out the reasons why your idea won't work before you've spent time, money, and other resources to implement it, or when you are up to your ears in investment and the undesired reality is staring you in the face, or staring a remote stakeholder in the face? If we attempt to identify the effects of our ideas before we set about implementing them, we can add elements to our ideas (additional injections) that will block the effects that would be unwanted. At a minimum, we can make a fully conscious decision to live with those effects before we move along the path of being confronted by them. Yes, there are many things we cannot predict. There are many that we can predict, though, if we only take the time.

If you are examining one of your own ideas, it may be very difficult (to put it mildly) to predict that anything even remotely undesirable could emerge from your great idea. Perhaps you'd like to ask a colleague to poke holes in your idea for you. If not, that's OK. All I'm asking you to do here is think about it for a moment and jot down any objections that might possibly come to mind. Don't worry, the process will bring you back to the same questions again at step 2b.

When you are ready to ask someone else for an opinion (now and/or in step 2b), seek out at least one person who you know will challenge your idea, and ask him or her to tell you what bad things can happen as a result of implementing it. Then, don't become defensive, don't argue, and don't discount what he or she tells you! If you need clarity, ask for it, but don't disagree. You will then have the opportunity to translate what the person has told you in terms of the categories of legitimate reservation.

Once you've got your list, make sure to **verbalize the concerns as entities, stated in the present tense**.

- At the printed circuit board company, the Vice President of Engineering feared that the demand for quick-turn prototype orders would swamp the design engineers. The Vice President of Manufacturing worried that salespeople would begin to offer three-day lead time on normal business, which would eventually erode the perceived value of quick-turn prototypes. This was later verbalized as, *The perceived value of quick-turn prototypes is eroding*. Other "Yes, but's..." centered more on reasons why this idea would be difficult to implement. These included things like, *Yeah, but we don't have the mechanism in place to quote jobs like this. Our current order entry process consumes three days in and of itself! How do we keep manufacturing capacity available for the quick-turn business, and not be 100% consumed with our regular work?* The management team wrote these down separately, and agreed to use a prerequisite tree to address any that remained after they completed the future reality tree.

- Objections to the "flat tax" included beliefs that it wouldn't be a fair tax for some citizens and that type of tax program would not generate the amount of income the United States needs to collect through income taxes. I also suspect that many employees of the Internal Revenue Service were concerned about job security. Take a moment now to translate these concerns into entity statements?

____

____

____

____

____

____

2. **Describe the effect–cause–effect relationships.** Now it is time to develop an understanding of how all these entities of the future — the injection, its objectives, and its potential undesirable effects — will be related to each other through cause and effect. In this part of the process, you will use sufficient cause thinking to uncover the assumptions being made, and test those assumptions for validity, using the categories of legitimate reservation. As a result, you will expose the need for additional entities and injections and will subsequently transform the initial idea (injection) into a much more robust solution.
   Here are some general guidelines:
   - **Verbalize all entities in present tense terms.** Stating the entities as if they exist in the present accomplishes a few things. First, it helps you to project yourself mentally into the future and visualize what you're thinking about. Second, when most of the entities on a tree contain the word "will" (as in *we will be more profitable*), the tree tends to be more cluttered and confusing both for the tree's creator and any of its readers.
   - **Do not use tentative types of phrasing such as *might, maybe,* or *possibly*** (as in *we might be more profitable*) in the future reality tree. As soon as you feel yourself reaching for the M-word ("maybe"), recognize that you have a causality reservation. Resolve the reservation by utilizing the categories for legitimate reservation. This will bring out the clarification you need about what else must exist in order to make your vision a reality and lead you toward adding the injections that will make it so. Don't settle for tentativeness.

a. **Using sufficient cause thinking, connect the injection to the objectives. You will use this step when:**
   - You are the champion or owner of the injection in order to clarify and expand it so that once implemented, it will cause the desired effects. You will then continue with step 2b.
   - You have been asked for (or simply want to give) your opinion of someone else's idea and are doubting that the idea will result in the desired effects. This step will help you to uncover and then communicate the assumptions you believe to be erroneous. It will also help you uncover any erroneous assumptions *you* are making as well.

You are now going to spend some time making sure that what you are planning to implement will actually achieve the desired effects. When you are the champion of an idea, you are intuitively claiming, **If** [injection], **then** [objectives]. Doesn't it make sense to you to first clarify what you want to implement and determine to the best of your ability that it will get you what you want? Doesn't it also make sense to do so *before* you worry about what additional consequences the injection may create, and *before* you worry about how you're going to go about actually implementing the idea?

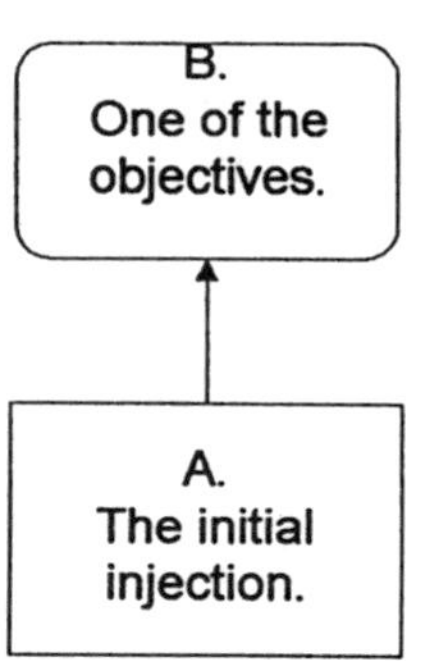

Figure 7.3

1. Select an objective that appears to be an obvious effect of the injection, and diagram the cause–effect relationship. Your diagram should look like Figure 7.3.
2. Scrutinize the relationship using the categories of legitimate reservation and modify as needed to solidify the connection. Your diagram may now look like Figure 7.4.
3. Look at the list of remaining objectives and compare to the entities on the tree you have constructed thus far. Select one that appears to be an obvious cause or effect of *any* of the entities already connected. Diagram that connection, scrutinize with the categories of legitimate reservation, and modify as needed to solidify the connection (Figure 7.5).
4. Repeat the step above, connecting *one entity* (*objective*) *at a time*, until all of the objectives from your objective list are connected to the tree.
5. At this point, you may decide to add additional desirable effects of the injection and other entities on the tree.

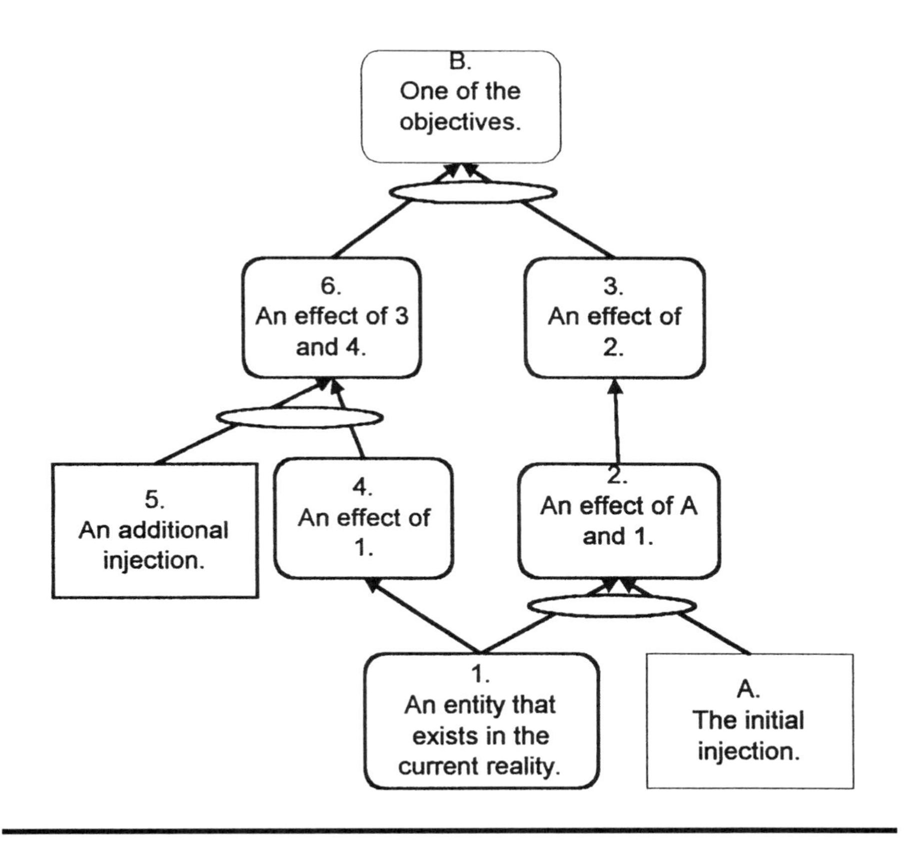

Figure 7.4

b. **Using sufficient cause thinking, seek and block potential undesirable consequences of the injection.** You will use this step when:

- You are the champion or owner of the injection, and you have completed step 2a.
- You have been asked for (or just want to give) your opinion of someone else's idea and believe that unpleasant things will happen once that idea is implemented. In other words, you had no trouble preparing a list of "yes, buts." This step will help you uncover and then communicate your concerns to your colleague in a clear and nonthreatening way. You may also identify some erroneous assumptions of your own.

Now that you have clarified the desired reality and what you have decided to implement in order to create it, it's time to develop an understanding of just how the solution might create

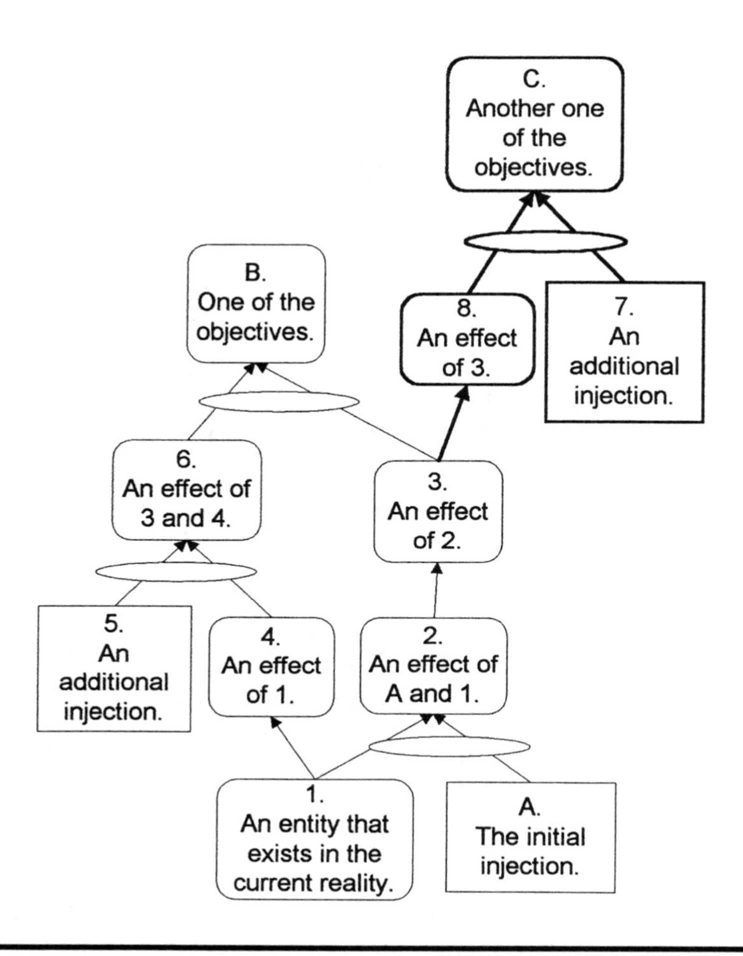

Figure 7.5

unwanted side effects. Yes, it's time to look those "yes, buts" straight in the eye and decide how you are going to prevent them from entering your magnificent future. Using the sufficient cause diagramming process combined with the categories of sufficient reservation, you will highlight opportunities to add injections that, if implemented, will prevent the unwanted effects from emerging.

Ask someone else for his or her scrutiny of your solution. Find *at least* one person who will not hesitate to challenge your ideas. Some principles apply here:

- The bigger the "butthead" the better for uncovering undesirable consequences.

The light that a man receiveth by counsel from another is drier and purer than that which cometh from his own understanding and judgement, which is ever infused and drenched in his affections and customs.

Francis Bacon, 1625

- The more stakeholders you involve, the better the solution.
- The more stakeholders you involve, the easier it will be to implement the changes.

Ask the person to tell you what bad things might happen as a result of the injections and their effects. I will repeat my earlier advice: Don't become defensive, don't argue, and don't discount what the person tells you! If you need clarity on how he or she reached the conclusions, ask for it, but don't disagree! Say thank you and write down their yes, buts. You will then have the opportunity to examine what you've been told as you proceed with the following steps.

1. Compare the list of potential undesirable consequences (con's) of the injection developed in step one of the tree. Given the injections, entities, and causal relationships developed in the tree, are any of them no longer relevant? Strike these from your list. Should any be added? Add these to your list. You are likely to answer both questions in the affirmative.
2. Select one of the "yes, buts" that appears to be an obvious effect of any of the entities on the tree (or of the initial injection, if you are at this step directly from step 1). Your diagram will resemble Figure 7.6.
3. Scrutinize the relationship using the categories of legitimate reservation, and modify as needed to solidify the connection. Your tree may now look like Figure 7.7.
4. You may decide to extend the "branch" so that the effects of the consequence are fully portrayed. You want the negative branch to illustrate the full "badness" of the effects you are predicting. Let's say that your spouse is really serious about this cruise idea, and your response is, "Yes, but I'll get seasick!" The negative branch that you prepare may look something like Figure 7.8.

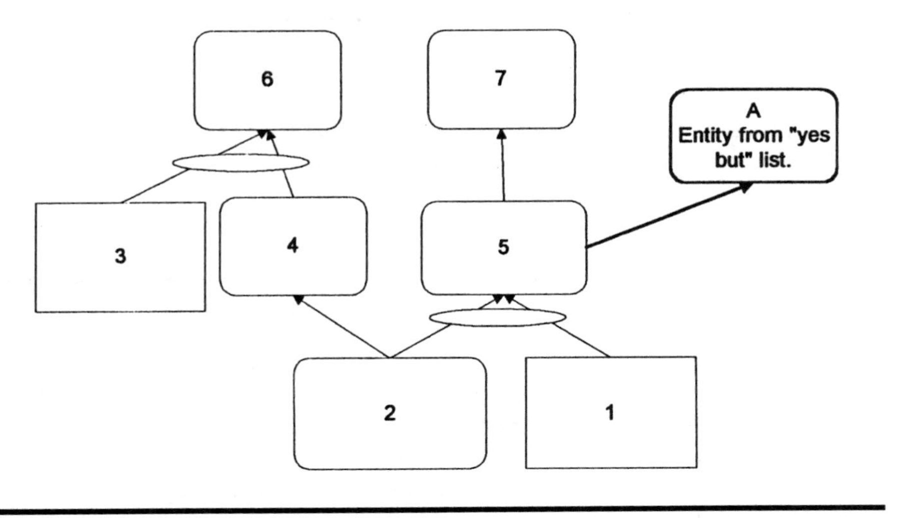

Figure 7.6

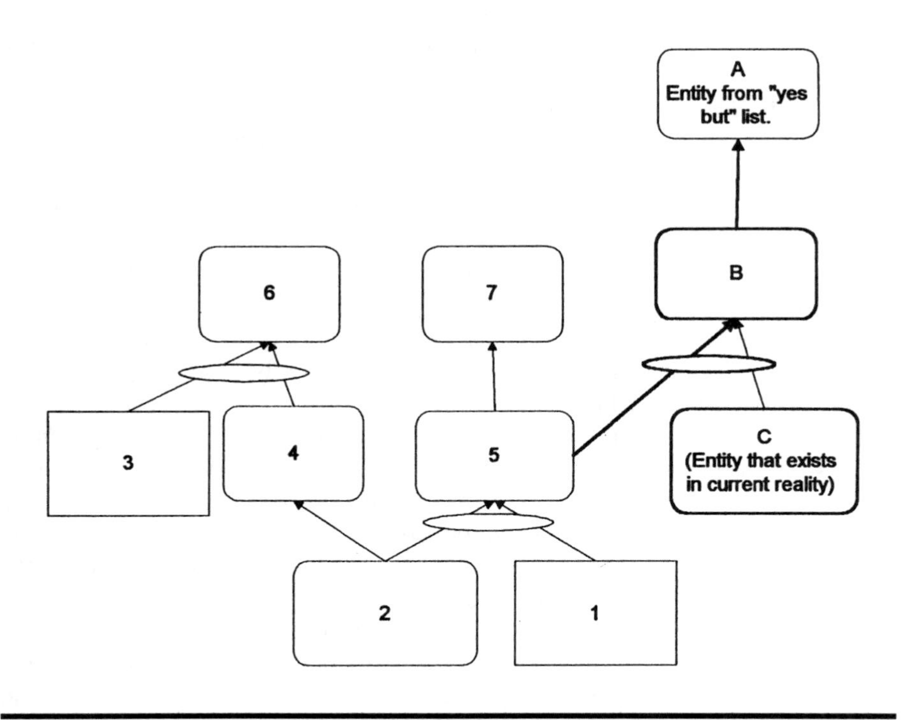

Figure 7.7

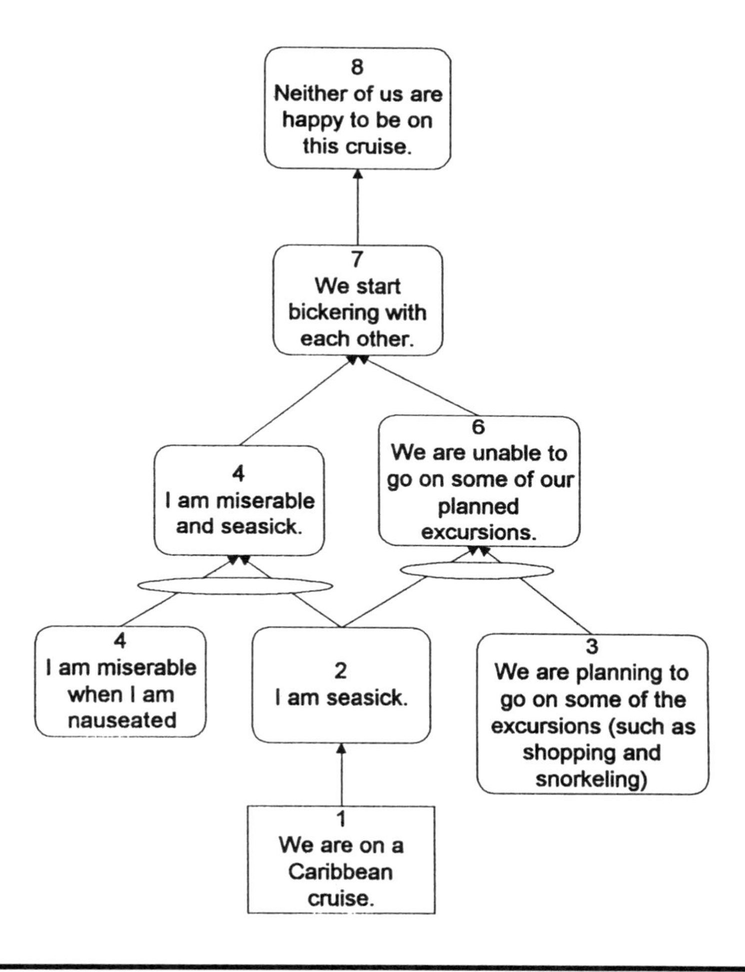

Figure 7.8

5. This is the fun part! Add an injection that will block the emergence of the undesired consequences, and recreate the tree to show the effects of that injection.
   a. As you read it from your starting point up, identify the first entity that is really undesirable on its own merit. It is at the arrow(s) feeding directly into this entity that you must place an injection. What, if you injected it into the future, would prevent the undesirable effect from emerging? Remembering once again that we are dealing with assumptions — those little gremlins that lurk beneath the arrows. I recommend that you do at least a mini-brain-

storming session. Generate as many ideas as you can, and then select the one that most appeals to you. Some possibilities for our cruise example might include:

- Take anti-motion sickness medication.
- Instead of a cruise, go on a land-based vacation (like Club Med) that has all of the amenities of a cruise, except the big boat.
- Rendezvous with your spouse at each port. Your spouse cruises there, you fly.

- How many more ideas can you think of?
    - ______________________________
    - ______________________________

Take a moment to identify which arrow each idea is aimed at, the assumption it breaks, and the reservation that was likely used to surface the assumption.

b. Recreate this portion of the tree to reflect the newly inserted injection and the future you predict as a result of its implementation. The previously predicted undesirable effects should not be on this diagram (Figure 7.9). Instead, you should see desirable (or at least non-undesirable) entities in their place.

6. It is quite possible that the decisions you have just made, as reflected in the modifications you have just made to your tree, are sufficient to deflect one or more of the remaining yes, buts on your list. If so, by all means strike them from your list, or build out the new branch to reach the desirable entities that replace them.
7. Repeat steps 2b1 through 2b6 until you have resolved all of the undesirable consequences you are able to predict.

3. **Enhance the solution.** If the tree you have just created reflects a solution you want to be long-lasting and self-regenerative, then you have just a couple more steps.
    a. **Predict additional effects**. Take a look at the entities on the tree and ask yourself what else they will cause. Using the categories of legitimate reservation, add the appropriate entities and arrows to the tree. This step often reveals that our solution may be much more far reaching than what is reflected in the diagram thus far.
    b. **Add reinforcing loops**. When creating the future reality tree for the purpose of designing the future environment of your system, this final step is very important. It defines the key means

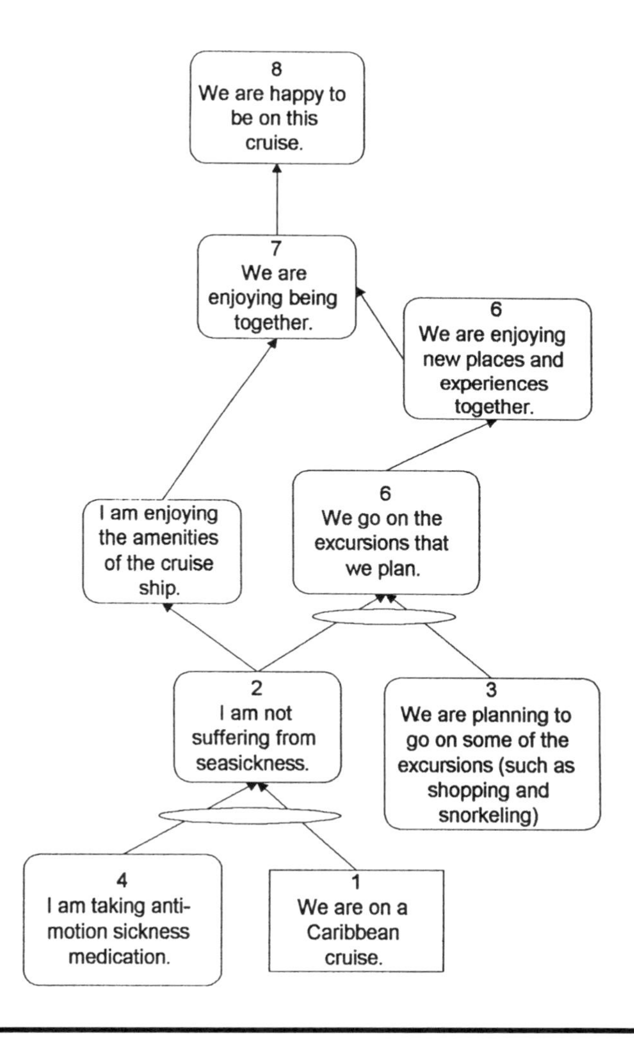

Figure 7.9

by which the system will sustain itself in ways that it will keep getting better. The bigger the loop, the stronger the systemic reinforcement.

- Examine the entities that are at the "top" of the tree. "Top" refers to those entities that don't have any arrows pointing out of them. Try to connect them to entities toward the bottom of the tree. For example, let's say your future reality tree looks like Figure 7.10.

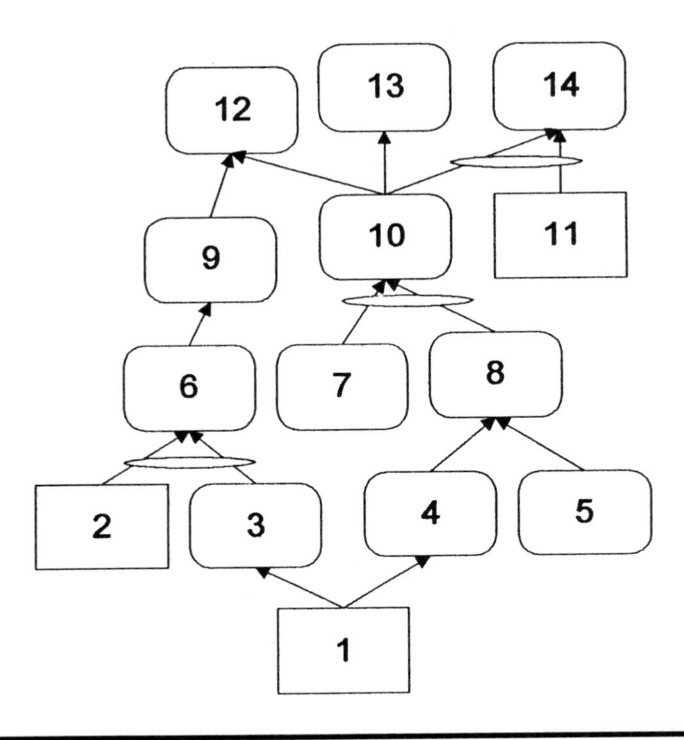

Figure 7.10

Your intuition tells you that Entity 13 is a cause for Entity 9. After using the categories of legitimate reservation, the connection is articulated by showing that it is actually Entities 13 and 14 that cause a new Entity 15. Entity 15 is actually a cause for Entity 9. You ask yourself if Entity 15 might be a cause for one of the lower entities in the tree.

- *If Entity 15, then Entity 3? No.*
- *If Entity 15, then Entity 4? No.*
- *If Entity 15, then Entity 5? Hmmn, now that seems real. Yes!*

Again you apply the categories of legitimate reservation. Because insufficiency points out that unless another condition is present, Entity 15 is insufficient to cause Entity 5. That condition does not exist, and it is not among any of the other entities in the future reality tree. Therefore, you add it — Injection 16. Now you are able to connect Entity 15 and Entity 16 to form a cause for Entity 5, creating a nice reinforcing loop (Figure 7.11).

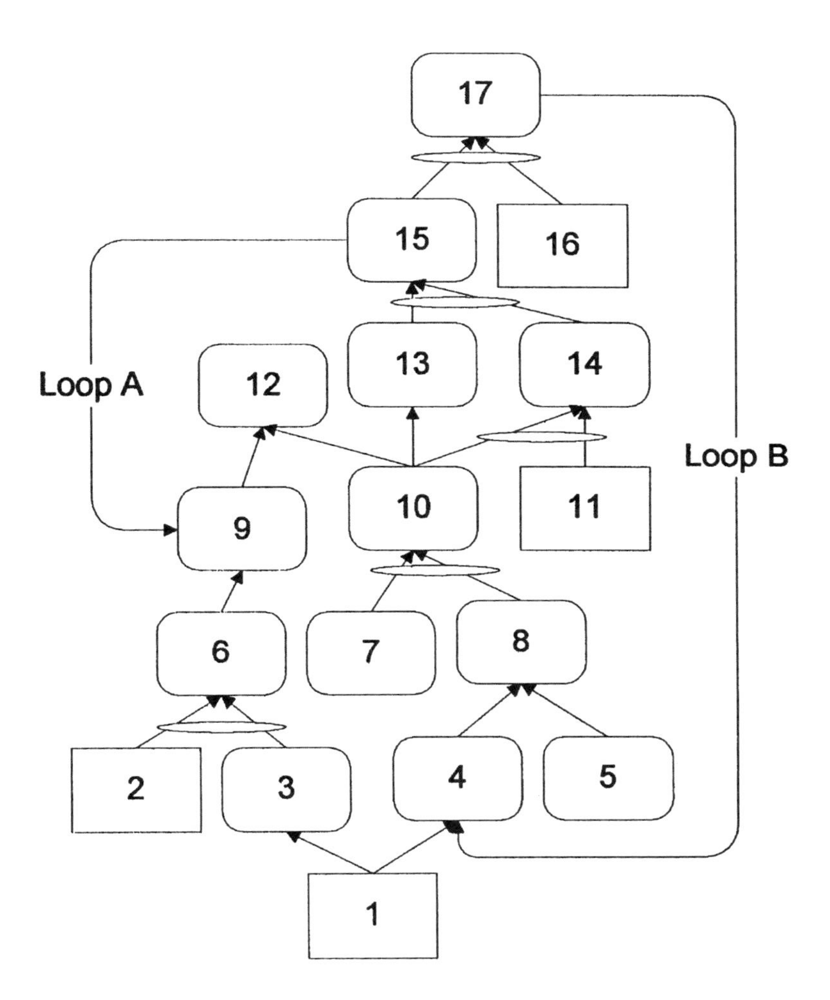

Figure 7.11

c. **Review the tree.** Read the tree back to yourself and confirm that it reflects the future you desire. Make sure the injections are things you are motivated to see through to implementation.
d. **Do something with it!** What you want is to be able to do a current reality tree one day in the future and have it look dangerously close to the future reality tree you have just completed. It's time to turn your attention toward implementation, and do what's necessary to make those injections happen.

- Some injections may be "no brainers" to implement. You know exactly what actions need to be taken, and there is nothing stopping those actions from happening other than just doing it. For these injections, "just do it."
- Some injections may appear almost as easy to implement, but you want to put some thought into just how you're going to accomplish it. You have an idea on how to get started, or perhaps can envision a series of steps, but you know that thinking through a plan will provide a much greater degree of assurance that the implementation will be successful. For these injections, use the **transition tree** to create your action plan. (See Chapter 6.)
- Some injections are faced with nontrivial obstacles. These are the injections that you look at and think, *How in the world are we really going to accomplish this?* It is not clear how you are going to get from here (the current reality) to there (the future reality in which the injection is implemented). For these injections, use the **prerequisite tree**. (See Chapter 10.)
- Your future reality tree will likely contain several injections. (If your tree has only one injection, I strongly suggest you revisit step two.) The implementation of those injections may very well involve coordinating the efforts of many people. It may involve allocating time and money. Sometimes, doing the easy stuff first simply lengthens the lead time to accomplish the global objective of getting all the injections implemented. In such cases, use the **prerequisite tree** to define the order in which to implement all of the injections. This will help to ensure that your resources are focused on doing what's important from a global objective perspective.

The following are examples of real future reality trees done by real people. They are unedited except for spelling. As such, I guarantee that they are not perfect. I have no doubt that as you read through them, you will have reservations. You will question arrows and entities, you will want to add entities (as described in step three), and no doubt you will have plenty of yes, buts.

The important thing is that the people who prepared these trees did so to the point at which they felt comfortable and confident in moving forward toward action. They put some quality thinking into the decisions they were trying to make. The thinking represents the intuition and care of every person involved in preparing the tree, and it left each of these groups with a document that they were able to refer to as time went on, checking their emerging realities with what they had predicted. They learned much in the process.

Every person who looks at a tree usually has something to add. I hope that you use this opportunity to practice the TP skills as your reservations emerge. Write down your thoughts, and then examine them with the categories of legitimate reservation in mind, and by studying the process itself. At what step would you have added whatever it is you would like to add?

Case One: Display Manufacturing

This illustrates an example of how one business used a future reality tree to create a solution to a core problem that was defined with a current reality tree.

In Chapter 8 (Current Reality Tree), we will travel with Display Manufacturing on their journey through the development of a current reality tree describing one of their market segments. Their subsequent future reality tree described the changes they decided to make to their product offering in order to alleviate many of their customers' problems. The purpose, of course, was ultimately to increase sales and profits for Display Manufacturing.

They began by defining the objectives. For each reselected pertinent undesirable entity on the current reality tree, the group answered, "*What will we replace this with?* Table 7.1 portrays the pertinent undesirable entities on the left, and the corresponding objectives for the future reality tree on the right.

They next decided on an initial injection. You may recall that the core problem they selected was, *Our product and service are designed for people with more than minimum skills.* They then spent some time brainstorming possible injections that were focused on eliminating the core problem. Here's a partial list of the potential injections generated in that session:

- Ship the racks assembled
- Ship a person with every rack
- Display Manufacturing coordinates rack setup, assembly, installation, and scheduling
- Design a simple rack
- Signage that explains exactly what the display is
- Send the rack with the product
- Idiot-proof the package

Table 7.1	
Pertinent Undesirable Entity	***Future Reality Tree Objective***
1. Display installation process is difficult for the customer.	The display installation process is pretty easy for store personnel.
2. Stores have a hard time receiving displays smoothly.	Stores are easily receiving our displays.
3. Installations are often a logistical nightmare.	Installations of our displays are going smoothly.
4. Amount of revenue generated by display doesn't justify costs associated with it.	Amount of revenue generated by display far outweighs the costs associated with it.
5. Installed costs are too high from the customers' perspective.	Installed costs are quite reasonable from the perspectives of the customer, the store, and Display Manufacturing.
6. Customers question the value of the display.	Customers rarely question the value of the display.
7. It's difficult for the manufacturers (our customers) to differentiate themselves.	The manufacturers are much better differentiated by the displays.
8. Displays aren't causing consumers to buy what they (displays) are meant to sell.	Displays are causing more and more consumers to buy what they're meant to sell.

- Call or fax to notify the store of pending delivery
- Send video or CD with the rack
- Provide a help-line
- GIVE them the display, customer pays "royalty"
- Talking rack

The injection they decided to start with was: *Display Manufacturing ensures that the display and its package are "idiot-proof" (shipped preassembled, labeled for easy receipt, preshipment notification, etc., based on the needs of the customer and store)*. Figure 7.12 illustrates the future reality tree they created.

Display Manufacturing had a nice list of obstacles to contend with in order to get to the point of actually offering their idea to the marketplace. They utilized the prerequisite tree process to define how they were going to overcome the obstacles. That process also helped them block some negative branches that they didn't articulate when they prepared the future reality tree.

Case Two: The Story of Max the Dog

This illustrates using the future reality tree to expand an initial idea (injection) into a unanimous family decision. While you will no doubt note that there are a good number of "technical flaws" in the tree (for example, we used the future tense in many of the entities), it is a good example of using the process in a situation where a decision must be made rather quickly.

In Chapter 4, I talked about the move my family made to another state. Before we moved, we had a pair of Golden Retrievers, named Max and Goldie. We gave Max and Goldie to our friend JR, who had a great back yard and two young boys to play with them. Within a month after our move, JR called us to say that there was no way he could keep Max. Goldie was fine, but both of them were too much. Max didn't seem to like JR's sons, and the dog's behavior had JR concerned. Before giving Max away or selling him, JR wanted to let us know. One of the choices we had was to take Max back. Our emotions told us to take Max back. We decided to use the future reality tree to help us make the decision.

Initial Injection: We take Max back.

We listed the reasons why taking Max back would be great. Jenn and Rachel (our daughters, who at the time were ages eleven and four, respectively) would be very happy. We would all feel more protected with a dog in the house to bark a big bark whenever someone came to the door. We remembered how good it felt when Max looked for a petting by leaning his head on our laps and looking up with his beautiful Golden Retriever "puppy eyes." Nothing like the unconditional love of a dog to warm the heart. We then sat down at the kitchen table, and as a family we drew this part of the tree (Figure 7.13), connecting the positive effects to the injection. I still have the piece of paper my daughter Jennifer wrote in very big letters, *"IF … THEN"* to remind herself how to read the words and arrows that I was documenting.

As you can see from Figure 7.13, we had an emotional stake in this animal. We even created a little tree that showed how guilty we'd feel if we let JR give him away to someone else. However, now it was time to look at the other side of the coin. I asked my family, "*What bad things will happen if we take him back?*" The girls, of course, could not think of a single problem. Danny and I, however, had no shortage of concerns. We even recalled a conversation we had several months earlier in which we heartily agreed: *No more pets.* So, treading somewhat lightly, we made our list of "yes, buts."

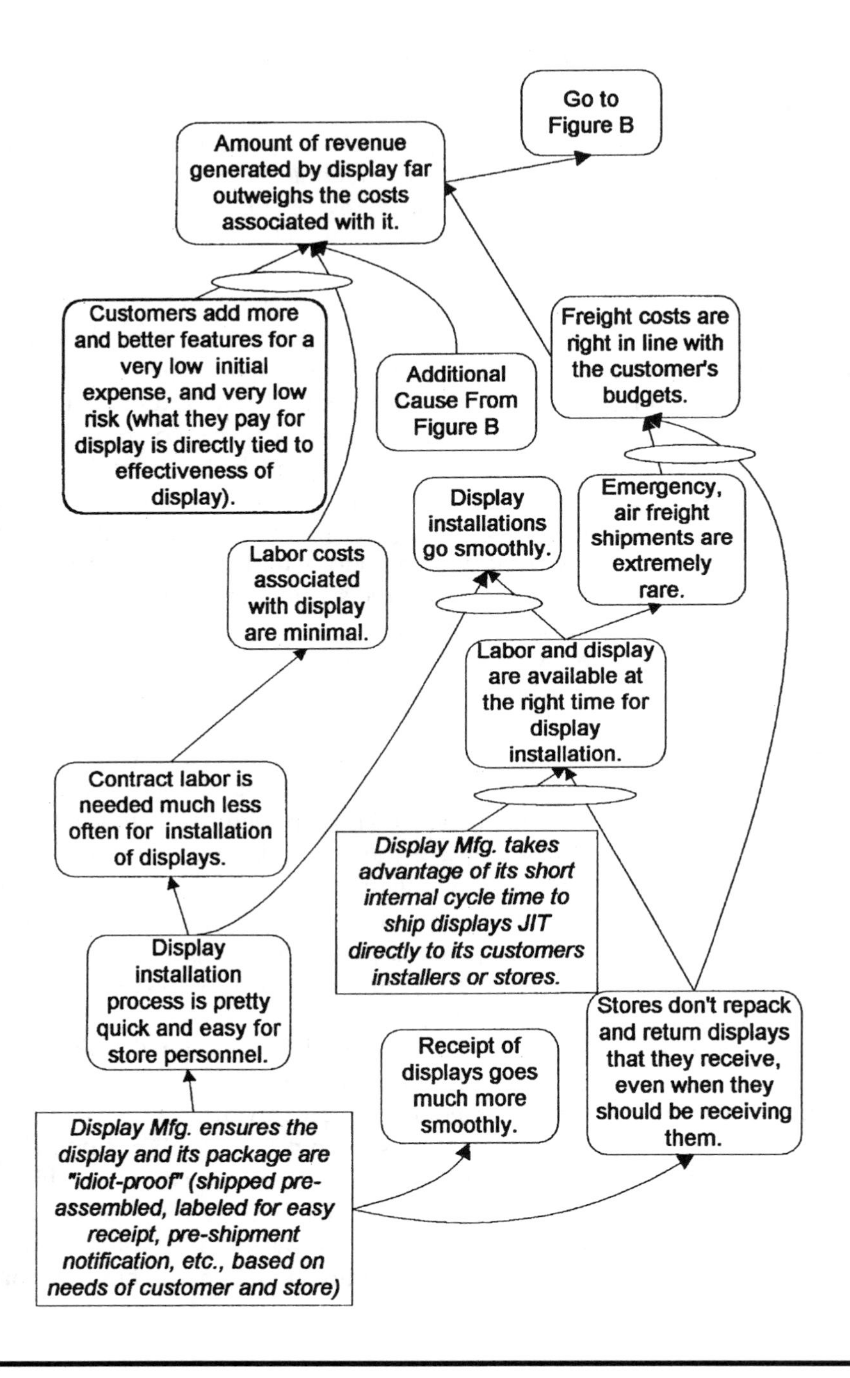
Go to Figure B
Amount of revenue generated by display far outweighs the costs associated with it.
Customers add more and better features for a very low initial expense, and very low risk (what they pay for display is directly tied to effectiveness of display).
Additional Cause From Figure B
Freight costs are right in line with the customer's budgets.
Display installations go smoothly.
Emergency, air freight shipments are extremely rare.
Labor costs associated with display are minimal.
Labor and display are available at the right time for display installation.
Contract labor is needed much less often for installation of displays.
Display Mfg. takes advantage of its short internal cycle time to ship displays JIT directly to its customers installers or stores.
Display installation process is pretty quick and easy for store personnel.
Receipt of displays goes much more smoothly.
Stores don't repack and return displays that they receive, even when they should be receiving them.
Display Mfg. ensures the display and its package are "idiot-proof" (shipped pre-assembled, labeled for easy receipt, pre-shipment notification, etc., based on needs of customer and store)

Figure 7.12a

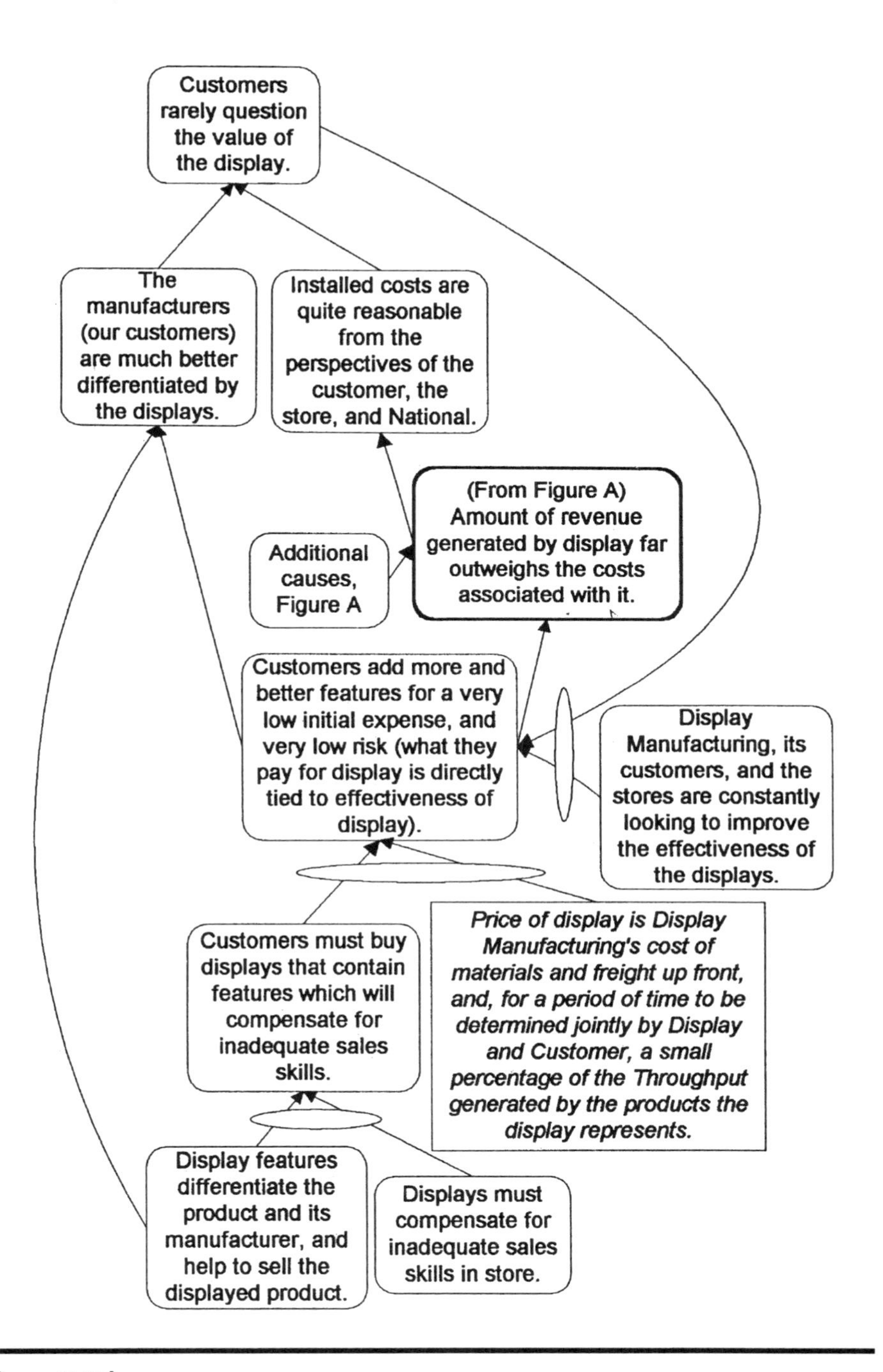
Customers rarely question the value of the display.
The manufacturers (our customers) are much better differentiated by the displays.
Installed costs are quite reasonable from the perspectives of the customer, the store, and National.
(From Figure A) Amount of revenue generated by display far outweighs the costs associated with it.
Additional causes, Figure A
Customers add more and better features for a very low initial expense, and very low risk (what they pay for display is directly tied to effectiveness of display).
Display Manufacturing, its customers, and the stores are constantly looking to improve the effectiveness of the displays.
Customers must buy displays that contain features which will compensate for inadequate sales skills.
Price of display is Display Manufacturing's cost of materials and freight up front, and, for a period of time to be determined jointly by Display and Customer, a small percentage of the Throughput generated by the products the display represents.
Display features differentiate the product and its manufacturer, and help to sell the displayed product.
Displays must compensate for inadequate sales skills in store.

Figure 7.12b

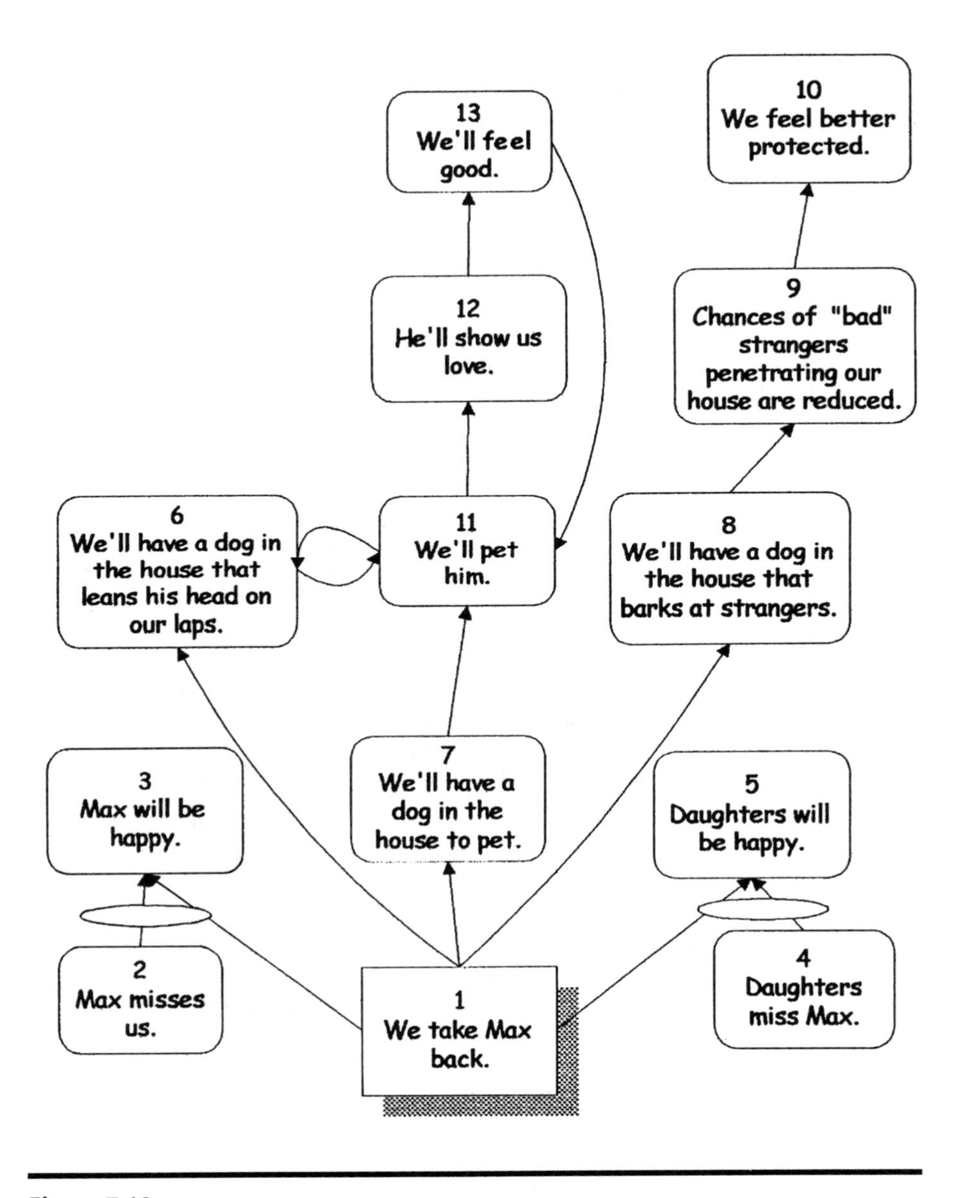

Figure 7.13

- **We (Danny and I) will get sick and tired of walking the dog.** Somebody would have to walk him a few times per day, and we knew that at some point in the near future, that chore would be exclusively ours.

- **We won't be flexible to take day, or longer, trips.** One of the exciting things about our move was the opportunity to take short trips to explore our new state. If we had this dog, how would we be able to do that on a whim?
- **Pieces of the house (that we were renting) may get damaged.** Max was known to chew on things like wooden molding and doors. When he and his sister Goldie were puppies, they even destroyed a section of carpeting in our hallway. We were renting our current house. It was one thing to have a dog that did damage to stuff we owned, and something much worse to have a dog that did damage to stuff we didn't own.
- **We'll spend money on fixing the house.** If he destroyed it, we'd certainly fix it, and that costs money.
- **Some of our belongings will likely get chewed up**. Max also really liked the taste of shoes.
- **The garbage can will always stink like dog poop.** Where else would you put the stuff you scooped? Add to this that the garbage can stayed in the garage until pickup day. The garage was attached to the house. Yuk.

As we talked through and wrote down each of the concerns, the girls would pipe in with a solution. We jotted down those ideas, too, but first we continued to create the rest of the tree by connecting the yes, buts to the injection (Figure 7.14).

Our bubble was definitely burst. Jennifer sadly said, *"Let's just not take him back."* While Danny and I certainly would have no qualms doing just that, we decided to test the process. *"Look,"* I said. *"Why don't we try to use the process to figure out how we can take Max back without suffering from all of these problems? It's worth a try, don't you think?"*

The four of us worked together at solving the negative branches, and the tree in Figure 7.15 resulted. The "**OK**" notations represent effects that we decided were neutral results, and we were willing to accept them into our future reality. For the most part, they represented additional expense we would incur as a family. We decided that if we could get Max back, and the rest of the tree could become our family's reality, we were more than willing to incur those expenses.

A week later Max was with us. We took him on a day-trip to the beach, where he had a great time playing with the girls and the waves. We signed up for obedience training, and soon had our first lessons. Then something happened that none of us would have ever predicted. While walking Max one morning, we came upon a neighbor who extended his hand to greet Max and allow him to sniff. Instead, Max bit him. Hard.

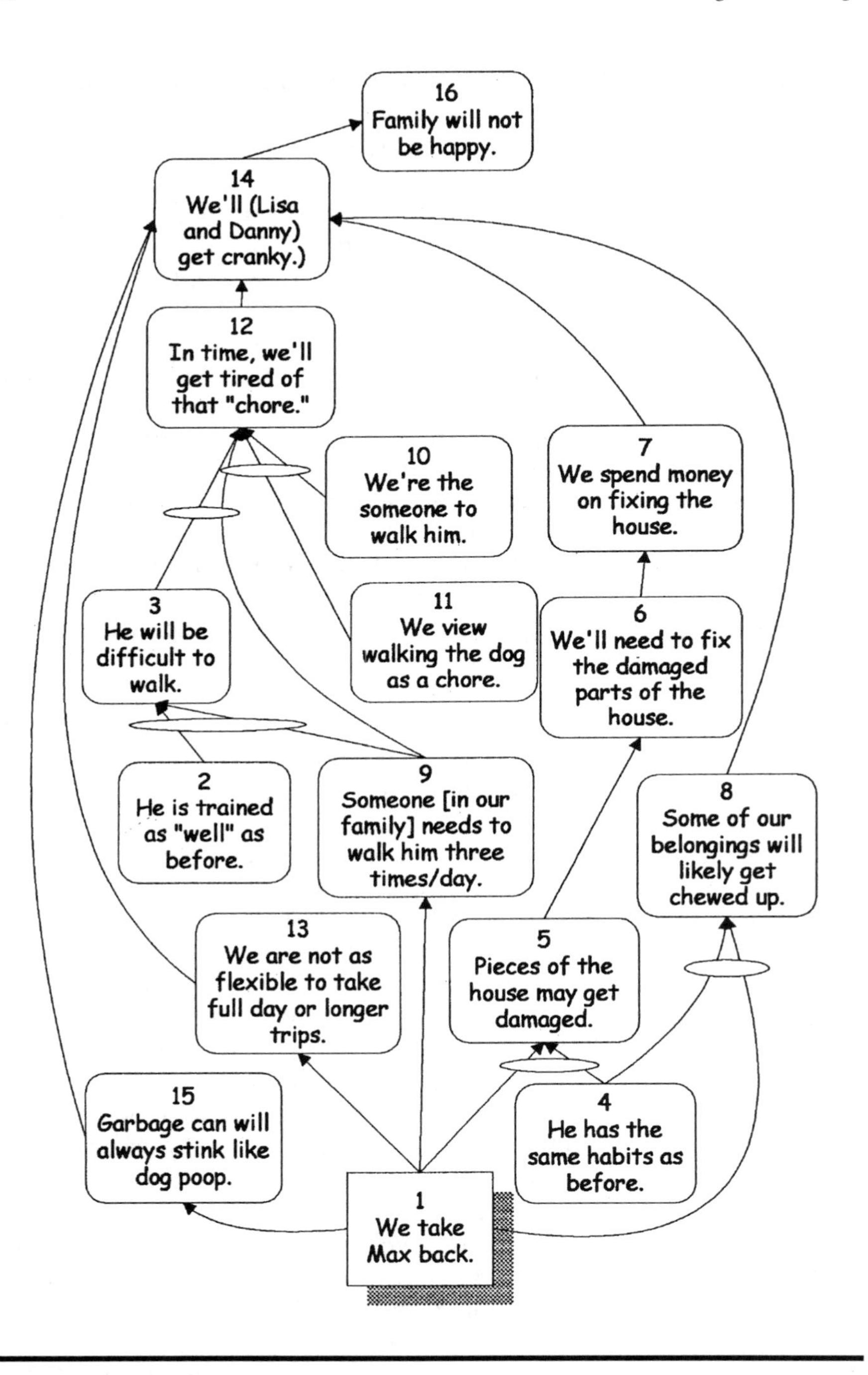
16
Family will not be happy.
14
We'll (Lisa and Danny) get cranky.)
12
In time, we'll get tired of that "chore."
10
We're the someone to walk him.
7
We spend money on fixing the house.
3
He will be difficult to walk.
11
We view walking the dog as a chore.
6
We'll need to fix the damaged parts of the house.
2
He is trained as "well" as before.
9
Someone [in our family] needs to walk him three times/day.
8
Some of our belongings will likely get chewed up.
13
We are not as flexible to take full day or longer trips.
5
Pieces of the house may get damaged.
15
Garbage can will always stink like dog poop.
4
He has the same habits as before.
1
We take Max back.

Figure 7.14

The neighbor needed several stitches. An hour after that, Max started to growl at Rachel. Of course, this was completely unacceptable. We called every animal and training group we, our family, and our friends could think of, but to no avail. Not a one could offer us help on bringing Max to a point that he could be trusted not to bite again. We had no choice but to send Max to the pound. After much research, we learned that an animal behaviorist could have predicted this sort of behavior in Max. However, such knowledge was nowhere in sight of our intuition nor that of anyone we knew at the time we prepared the tree. Does that mean our tree was faulty, or a waste of time? We certainly don't think so. We made the right decision with what we knew, and we never regretted it. Subsequently, we learned more than we ever thought we'd know about dog behavior. We also had a great opportunity to use the thinking process tools as a family and were able to reach a decision together, without arguing. The process itself brought us closer.

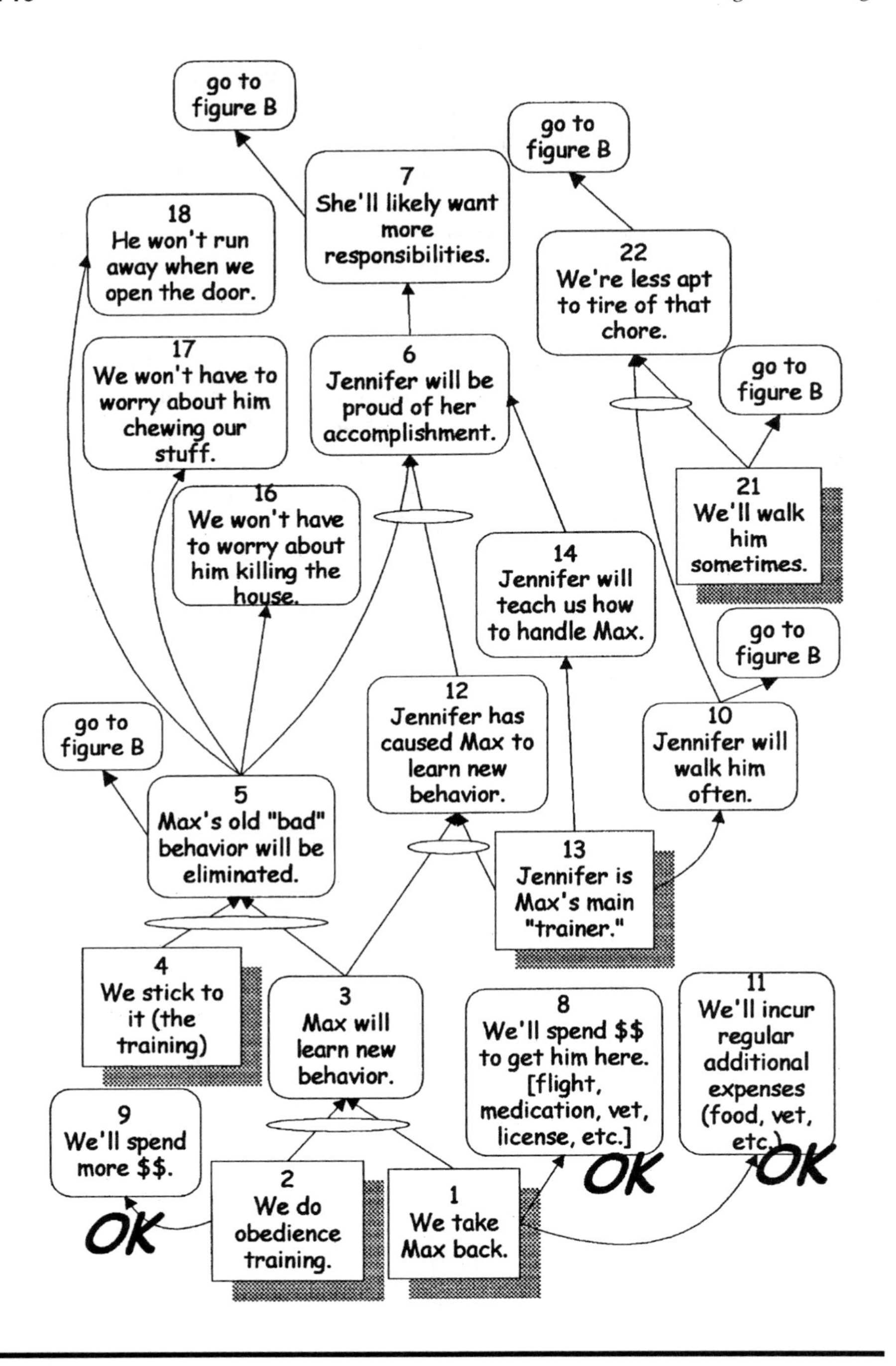
go to figure B
7 She'll likely want more responsibilities.
go to figure B
18 He won't run away when we open the door.
22 We're less apt to tire of that chore.
17 We won't have to worry about him chewing our stuff.
6 Jennifer will be proud of her accomplishment.
go to figure B
16 We won't have to worry about him killing the house.
21 We'll walk him sometimes.
14 Jennifer will teach us how to handle Max.
go to figure B
go to figure B
12 Jennifer has caused Max to learn new behavior.
10 Jennifer will walk him often.
5 Max's old "bad" behavior will be eliminated.
13 Jennifer is Max's main "trainer."
4 We stick to it (the training)
3 Max will learn new behavior.
8 We'll spend $$ to get him here. [flight, medication, vet, license, etc.]
11 We'll incur regular additional expenses (food, vet, etc.)
9 We'll spend more $$.
OK
2 We do obedience training.
1 We take Max back.
OK
OK

Figure 7.15a

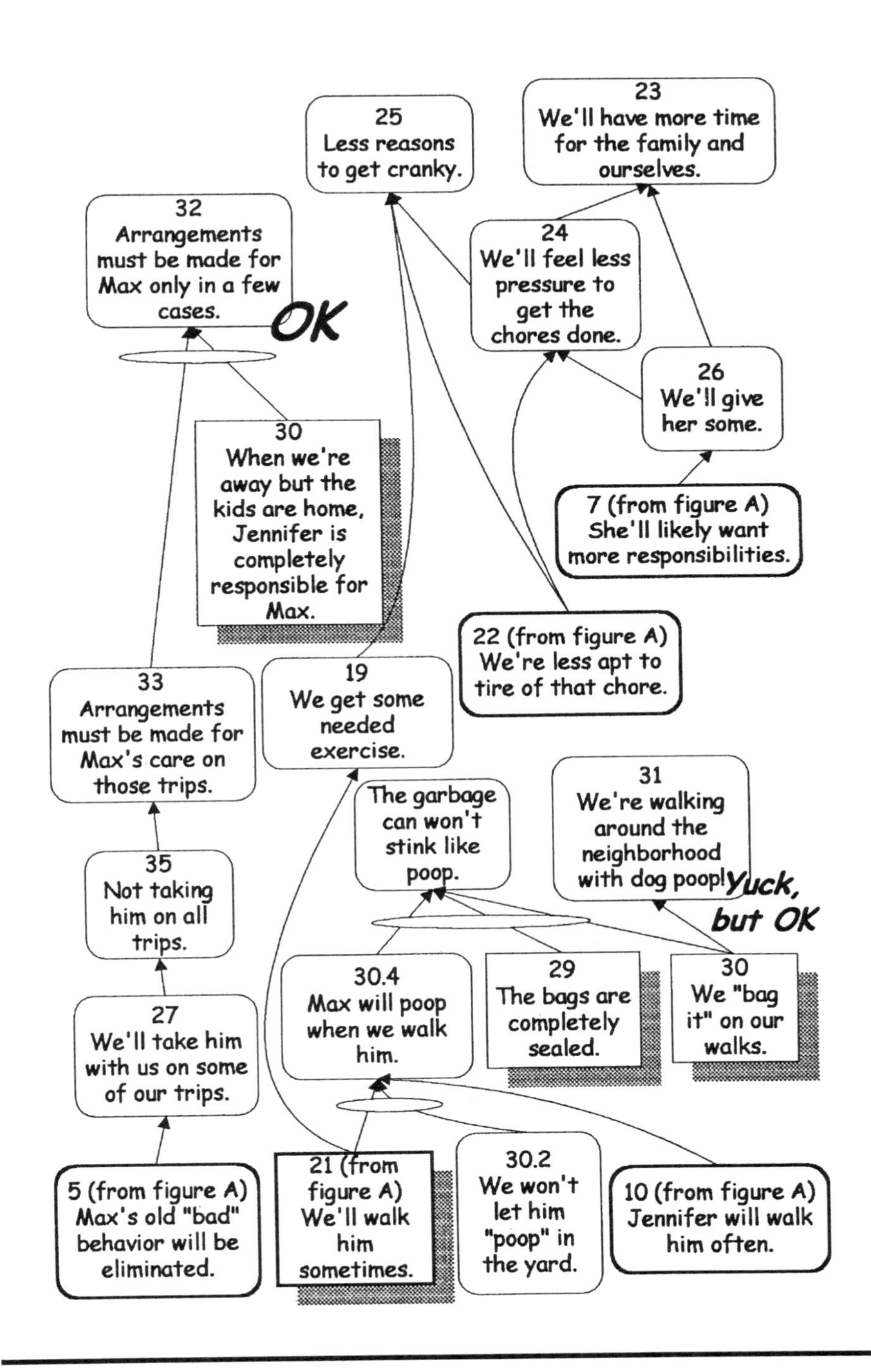
25
Less reasons to get cranky.
23
We'll have more time for the family and ourselves.
32
Arrangements must be made for Max only in a few cases.
OK
24
We'll feel less pressure to get the chores done.
26
We'll give her some.
30
When we're away but the kids are home, Jennifer is completely responsible for Max.
7 (from figure A)
She'll likely want more responsibilities.
22 (from figure A)
We're less apt to tire of that chore.
33
Arrangements must be made for Max's care on those trips.
19
We get some needed exercise.
The garbage can won't stink like poop.
31
We're walking around the neighborhood with dog poop!
Yuck, but OK
35
Not taking him on all trips.
30.4
Max will poop when we walk him.
29
The bags are completely sealed.
30
We "bag it" on our walks.
27
We'll take him with us on some of our trips.
5 (from figure A)
Max's old "bad" behavior will be eliminated.
21 (from figure A)
We'll walk him sometimes.
30.2
We won't let him "poop" in the yard.
10 (from figure A)
Jennifer will walk him often.

Figure 7.15b

Chapter 8

Current Reality Tree

> The simpler way to organize human endeavor requires a belief that the world is inherently orderly.
>
> Margaret Wheatley

Do you or your organization have any of the following symptoms?

- The list of problems you are facing is quite long.
- It is difficult to decide what to spend your time on, because there are so many projects screaming to get completed.
- You have the sense that you are spending most of your time in a swimming pool that has hundreds of ping pong balls floating on the surface, and your job is to hold them under the water with only your own two hands.
- It seems that chaos rules.
- You're trying to solve the same old problems with the same old solutions (though the solutions might be packaged a little bit nicer than they were last time you tried them).
- You are having a difficult time identifying the physical constraint of the system.
- You are having a difficult time getting clarity on just what the purpose of the system is.
- You want to understand the policy and/or paradigm constraints of the system and the far-reaching impact of those constraints on the performance of the system.

- You want to discover the core competency or key strength of the system.
- There is a need to understand the answer to the question, "Why should we even bother to change our existing system?"

Enter the current reality tree, the thinking process application tool that is used to pinpoint a core driver — a common cause for many effects. The most common use for the current reality tree is to identify a core problem, which can be thought of as the invisible constraint responsible for many of the system's current problems. Emerging uses of the tool are to identify core competencies and strengths, and to simply understand why change is necessary in a given environment.

Constructing a current reality tree is a process that leads to recognition of the behavioral patterns of the conditions existing in the reality of the system. It is a tool that allows us to see order even in the midst of chaos.

Reality is that which, when you stop believing in it, doesn't go away.

Philip K. Dick, 1985

Unlike other tools that search for root causes, most of the current reality tree's process steps do not focus on diving deeper and deeper into the issues. Rather, we are led through a routine of examining and documenting the cause–effect relationships that exist between and among conditions that are present in the system. Only in the final steps of the process, after the relationships are described, are we guided to discover the common causes and identify the core driver.

The current reality process contains six major steps:

1. Determine the scope of the analysis.
2. List between 5 and 10 pertinent entities.
3. Diagram the effect–cause–effect relationships that exist among the pertinent entities.
4. Review and revise for clarity and completeness.
5. Apply the "so what test."
6. Identify the core cause(s).

In the following pages, I describe each of these steps in detail.

Step 1. Determine the Scope

Scope can be as large or as small as you want. Limit the scope to something you care about, something you are willing to spend your own time analyzing, developing the solution, and working on its implementation. Limit the scope to something you know about so you can honestly say you have "intuition" into the reality of the subject. Remember, intuition comes from personal experience.

For instance, if you work for a large corporation and you are going to do the full analysis on your corporation, don't do it yourself! By involving people from other disciplines in the organization, the chances will be much better that your analysis will reflect the realities of the organization as a whole.

A few years ago, I used the current reality tree to try to understand what was causing an organization's problems. The "system" consisted of a franchiser and its franchisees. Among the issues they faced were: sales were low; franchisees deeply distrusted the franchiser's motives whenever the franchiser issued changes to the franchise policies; and the franchiser thought the franchisees were not pulling their weight, while at the same time the franchisees believed their suggestions on how to increase business fell on deaf ears. Well, by the time I finished the tree it was about 10 pages, in eight-point font. Only two of those pages were about the original system I was tasked to analyze. The remaining eight pages, as it turned out, were an analysis of "the nature of our society's win-lose paradigm." A great learning experience, but way off task.

Ask yourself (and answer!) the following questions before you start the tree, and you will go a long way toward avoiding paralysis by analysis disease.

1. What is the system you are analyzing?
   a. What are its boundaries?
   b. Can you picture the 40,000' view? (See Chapter 2.)
2. What is its purpose? Why does it exist?
3. What are the major measures of success? For instance, if you were analyzing a for-profit corporation, two of the measures of success would be net profit and return on net assets.
4. What will this current reality tree help you to understand?
   a. Are you going to identify a core problem?
   b. Are you going to determine a core strength?
   c. Are you going to determine a core driver of the entire system (what is it that drives the system's strengths *and* weaknesses)?

Example: Display Manufacturing

A small manufacturer of metal display racks (we will call Display Manufacturing) needed to increase sales. They wanted to do this in a way that increased their customers' perception of value, so that they could protect, or even increase, the prevailing prices of their products. They decided to utilize the current reality tree to identify areas which, if they targeted, would solve real problems for their customers, and thereby provide real value.

Display Manufacturing's racks are seen in tire stores, home improvement stores, and discount stores. Display Manufacturing sells its racks to the manufacturers of products that are sold in the stores. The supply chain is shown in Figure 8.1.

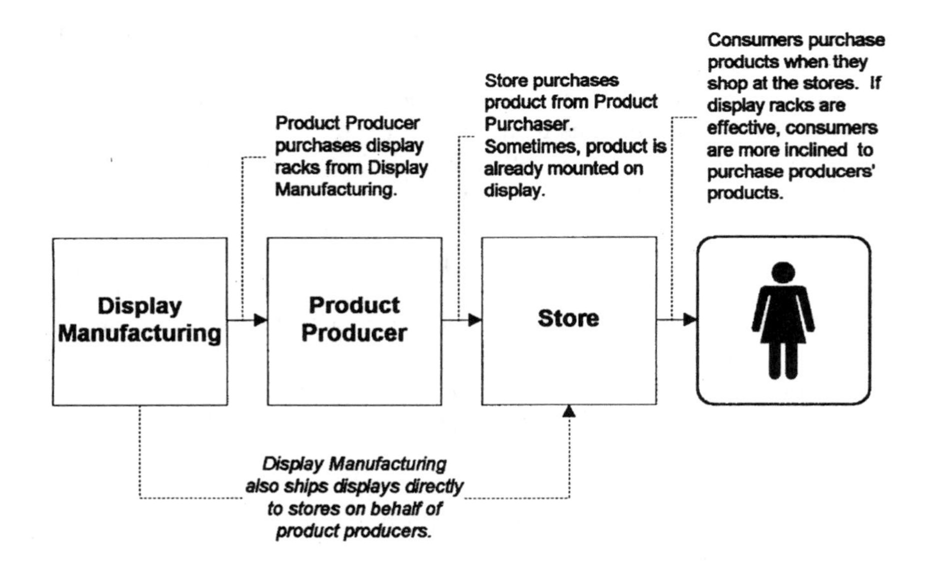

Figure 8.1

A team of people that represented various functions of Display Manufacturing's business was assembled. Included were manufacturing, accounting, engineering, sales, marketing, an employee who had previously been employed by a producer, and an employee who had previously been employed by a store. The Display Manufacturing team decided to look at its impact on the bottom line of its customers, the product producers. The place where this really happens is at the store itself. If the store sells more product, the producers sell more product. So the question Display Manufacturing was determined to answer was, "How can our product offering help the producers sell more product?" For added focus, the team selected a specific segment of the market — the system through which Display Manufacturing shipped displays directly to the stores on behalf of the producers.

Step 2. List Between 5 and 10 Pertinent Entities that Exist in the System, Relative to the Scope of the Analysis You Have Defined in Step One

Although it may be quite easy to generate a long list of pertinent entities (especially when you're focusing on problems), please limit this to no fewer than 5 and no more than 10. This initial list provides the starting point for your analysis. Later in the process you will be invited to add entities that are relevant. **Subject each of these pertinent entities to the entity existence and clarity reservations.** (See Chapter 4.)

1. When creating a current reality tree to determine a core problem, the pertinent entities are called undesirable effects (UDE). Identify them with your answers to the following:
   a. What evidence indicates that the system is misaligned with its purpose?
   b. In what ways is the system achieving fewer of its goals than it desires to achieve?
   c. What exists in the system that is preventing it from achieving more of its goals?

   Beware of the tendency to claim what I call *the lack of my preconceived solution* as a pertinent entity. For instance, you may be bothered that your company hasn't implemented the XYZ quality improvement program, or that you and your spouse have not set aside a predetermined time every week to spend together, or that your organization doesn't have high-performance work teams or a training program for new managers. None of these describes entities that exist. They describe solutions that don't exist. So what would the quality program change, if only it was implemented? Perhaps something like, *Our yield for product line ABC is 70%*. What would setting aside a predetermined time every week with your spouse do? Maybe it would address the entity, *My spouse and I rarely have time to spend together these days*. Try to get at the entities that do exist rather than holding on to solutions you have desired. Your analysis will be more robust and truthful. I guarantee that if you go into an analysis already confident that problems exist simply because your solution hasn't been implemented, sure enough, that's exactly what you'll prove to yourself. Remember to put on your "learning hat."

2. When creating the current reality tree to determine a core strength, you are looking for the good, strong aspects of the system. Thus, the pertinent entities will emerge with your answers to the following:
   a. In what ways is the system strongly aligned with its purpose?
   b. In what ways are the system's stakeholders (customers, employees, owners, suppliers, community, etc.) pleased?
   c. What are the system's strengths?
3. When creating the current reality tree to determine a core driver of the whole system, remember the old adage, "Your strongest strengths are also your weakest weaknesses." The pertinent entities will be answers to the questions in (a) and (b) above, selecting 5 problems and 5 strengths.
4. *Don't forget to subject each entity on your list to the entity existence and clarity reservations.*

Display Manufacturing (continued)

Because Display Manufacturing was looking for customer problems to solve, it was using the current reality tree to find a core problem. Specifically, they were going to identify the core problem of the system in which Display Manufacturing shipped displays directly to stores on behalf of their customers, the producers. Remembering that the purpose of the analysis was to ultimately determine how Display Manufacturing might change its product offering in order to alleviate the identified core problem, the initial list of pertinent entities was a list of problems experienced by their customers and stores as participants in this system. Their initial list of pertinent entities is shown in Table 8.1.

Step 3. Diagram the Effect–Cause–Effect Relationships that Exist Among the Pertinent Entities

A. Take a look at your list of pertinent entities. Identify two entities from your list that appear to be involved in a sufficient cause relationship and that meet one of the following two criteria. Then, diagram the relationship as in Figure 8.2.

Table 8.1

Initial	*After Entity Existence & Clarity Reservations*
Store personnel have no skills.	1. To keep their costs low, stores hire personnel with minimal skills.
Customers question the value of the display.	2. Customers question the value of the display.
Displays cost too much.	3. Installed costs are too high from the customers' perspective.
It's difficult for customers to differentiate themselves.	4. It's difficult for the stores to differentiate themselves.
	5. It's difficult for the manufacturers (our customers) to differentiate themselves.
Customers are constantly looking to reduce the costs of the displays.	6. Customers are constantly looking to reduce the costs of the displays.
Installations are a mess.	7. Installations are often a logistical nightmare.
Display installation is difficult for the customer.	8. Display installation is difficult for the customer.

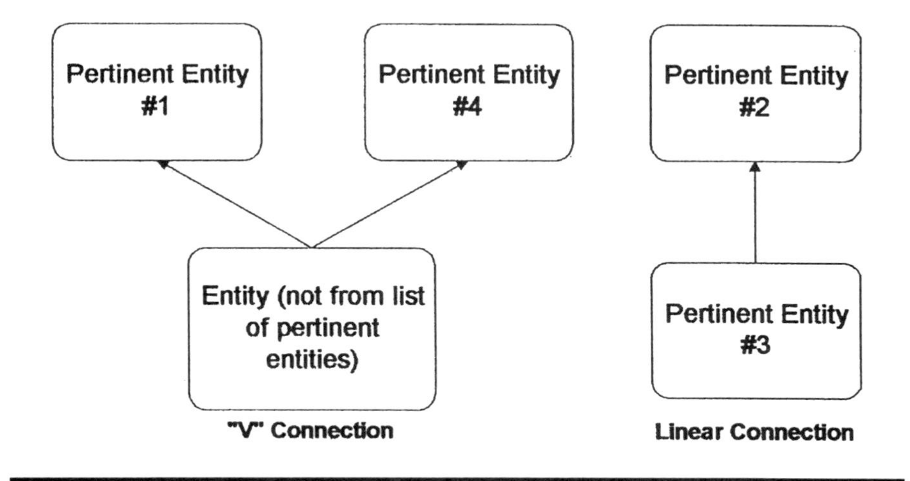

Figure 8.2

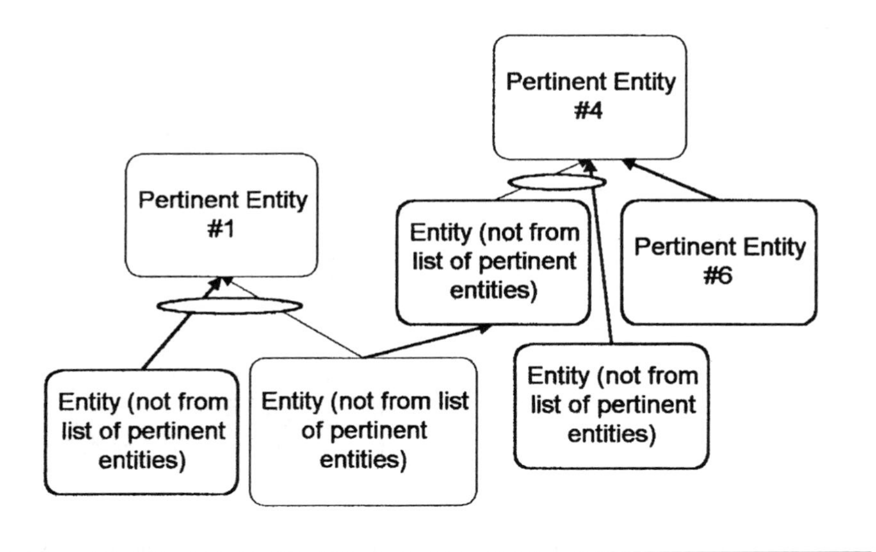

Figure 8.3

1. A relationship in which you identify a common cause for two of the pertinent entities is referred to as a "V Connection."
2. A relationship in which you identify one of the pertinent entities as a cause for another of the pertinent entities is called a "Linear Connection."

B. Scrutinize this first connection using the categories of legitimate reservation, (see Chapter 4) modifying your tree as your use of the categories prescribes. For instance, after applying the categories of legitimate reservation, the diagram in which you first identified a "V" connection as illustrated in Figure 8.2 might now look like Figure 8.3.

C. Look at the remaining list of pertinent entities and compare them to the entities on the tree you have constructed thus far. Select one of the remaining pertinent entities that appears to be involved in a sufficient cause relationship with any one of the entities on the tree, and draw the connection, as in Figure 8.4.

D. Repeat steps (3B) and (3C) above — *connecting one entity at a time* — until all of the entities from your pertinent entity list are connected to the tree. This means that each pertinent entity has an arrow pointing to and/or from it.

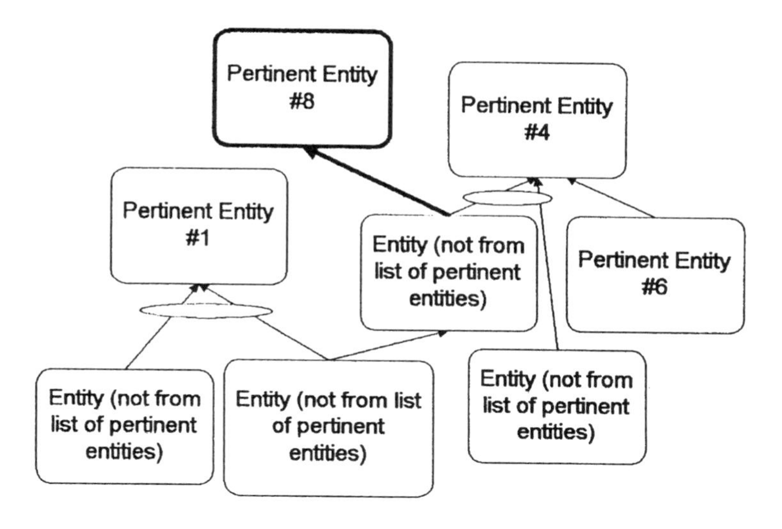

Figure 8.4

Process Tips

- Concentrate on simply connecting each of the pertinent entities, **one at a time**. You will have an opportunity in subsequent steps to expand the tree beyond this limitation.
- Avoid the temptation to keep "diving down" into the causes for the entities that have no arrows pointing into them.
- Do not add to your initial list of pertinent entities. At this point in the process, no entity is needed unless it is required to connect a pertinent entity to the tree, or it has been added as a result of applying the categories of legitimate reservations.

Display Manufacturing (continued)

Before we take a look at Display Manufacturing's tree, I want to warn you. It's not perfect! This is an example of a real tree created by real people. The objective was not to have a *perfect to the nth degree* tree, but rather to get to a point of understanding the current reality. This was not an academic exercise, but a live application of the tool. Therefore, I suspect that, as you read through the remaining elements of this example,

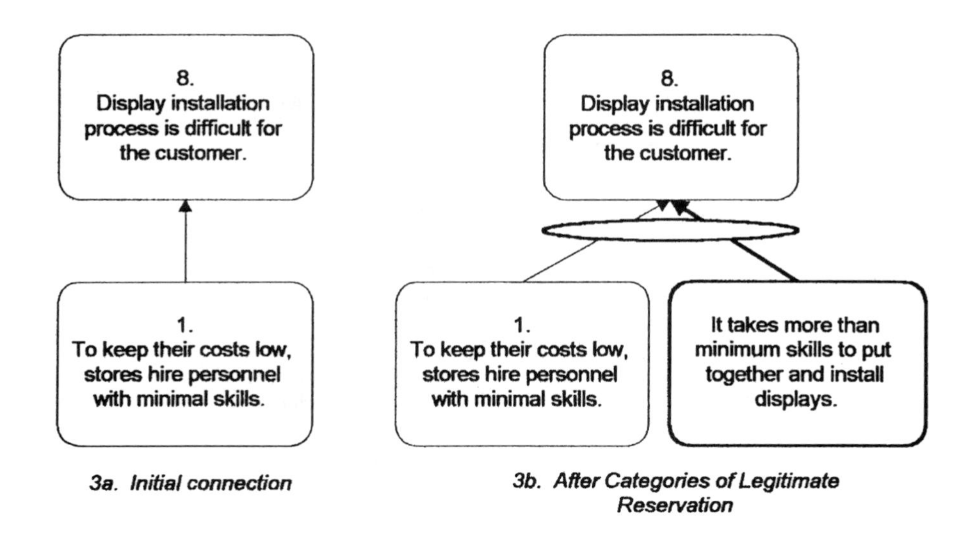

Figure 8.5

you will have several reservations, some of which will be quite valid. Recognize that those reservations don't mean that Display Manufacturing's tree is bad. It just means that had you been there raising the reservations, the clarity of the tree would have been enhanced. As you read through the evolving example, I encourage you to take a pencil and, using the categories of legitimate reservation, note your reservations. Then, try to answer your reservations by asking yourself in the following frame of reference: **Assume that the connection is, in fact, valid. What else would need to be in this space on the tree in order for me to accept it as such?**

This will give you excellent practice in using the categories of legitimate reservation, and thus enhance your sufficient cause thinking skills. It will also help you increase your ability to "put yourself in someone else's shoes" when listening.

It appeared to the team that entity *#8* (*Display installation process is difficult for the customer*) was an obvious effect of entity #1 (*To keep their costs low, stores hire personnel with minimal skills.*) The insufficient cause reservation led them to clarify the relationship further, as illustrated in Figure 8.5.

As the team looked at the rest of the list, they found it difficult to connect a third pertinent entity to the tree. Some seemed to be obviously connected to each other, but none seemed to be closely connected to the cluster that they started. So, they methodically went through verbalizing a hypothesized cause–effect relationship between each of the three entities on the cluster with each of the remaining entities on the list. "**If** *to keep their costs low, stores hire personnel with minimal skills,* **then** *customers question the value of the display.*" Does it make sense? No. Ok, move on to the next. "**If** *to keep their costs low, stores hire personnel with minimal skills,* **then** *installed costs are too high from the customers' perspective.*" Does it make sense? No. OK, move on to the next, and so on. Ultimately, an intuitive bell rang when they tried, "**If** *display installation process is difficult for the customer,* **then** *installed costs are too high from the customers' perspective.*" (See Figures 8.6 and 8.7.)

Step 4. Review and Revise for Clarity and Completeness

A. Now that all of the pertinent entities are connected, take the opportunity to expand the tree to gain a more complete understanding of the current reality. What are the effects of the entities, or combinations of entities, on the tree? What other aspects of the system are affected? What other stakeholders of the system are affected?
   - If your current reality tree is for the purpose of identifying a core problem, look for additional undesirable effects of the entities on your tree.
   - If the purpose of the tree is to identify a core strength or competency, look for additional desirable effects of the entities on your tree.
   - If your tree is to determine a general core driver of the system, look for good as well as problematic effects of the entities on your tree.

B. Look for missing connections that exist among the entities on the tree.
   - Of course, expand the tree by diagramming these relationships and subjecting them to the categories of legitimate reservation. *Notice, I have still not said that you should "dive down." Even at this stage of the process, it remains an unnecessary task.*

C. Read the tree back to yourself. It's easiest to follow if you begin at the lowest entry point and go up from their. Many people find it helpful to use the terminology "if...then" to designate "cause–effect" when reading a tree. "If [cause], then [effect]. If you can't read the "If [cause], then [effect]" statement in a free-flowing

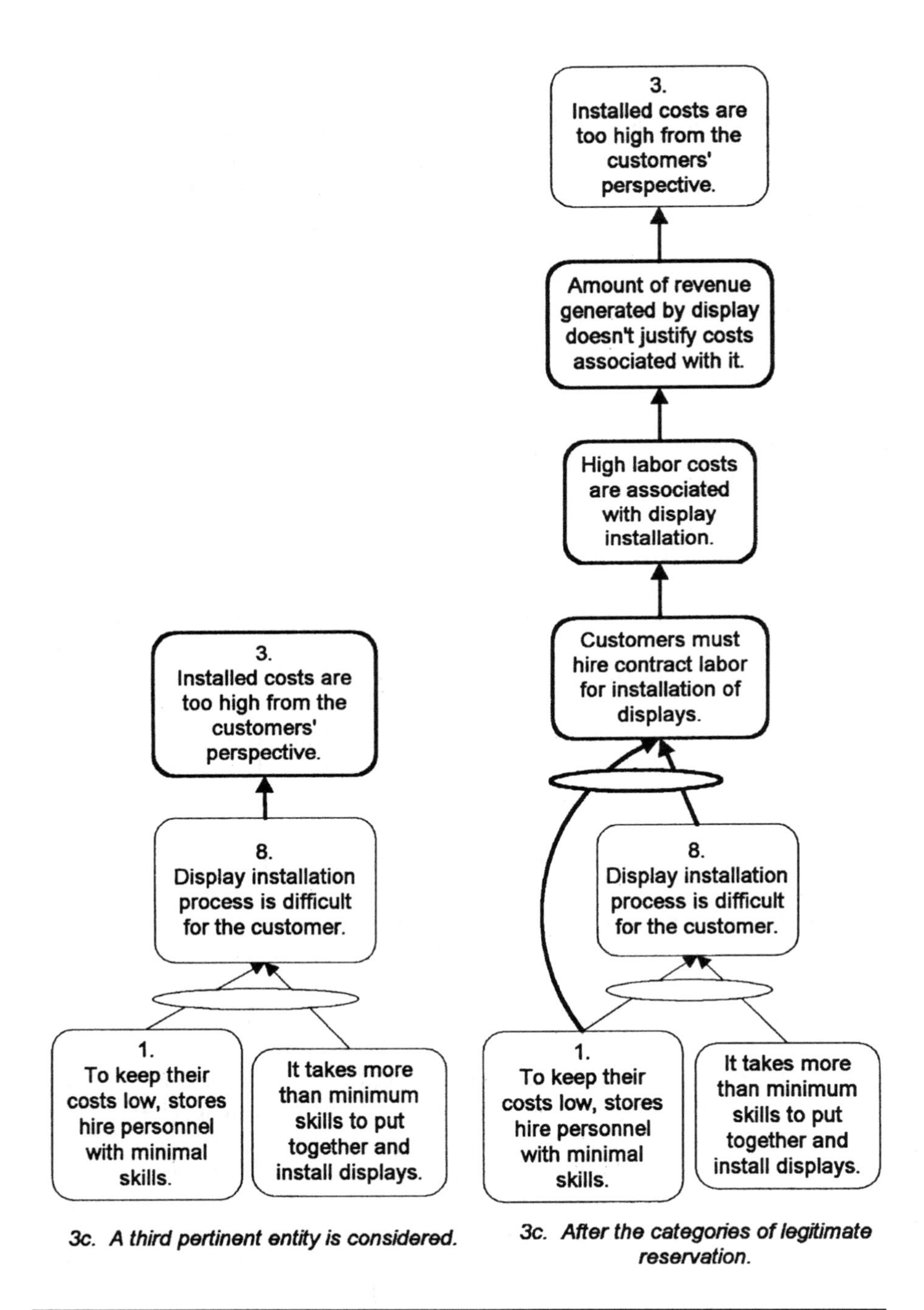

3c. A third pertinent entity is considered.

3c. After the categories of legitimate reservation.

Figure 8.6

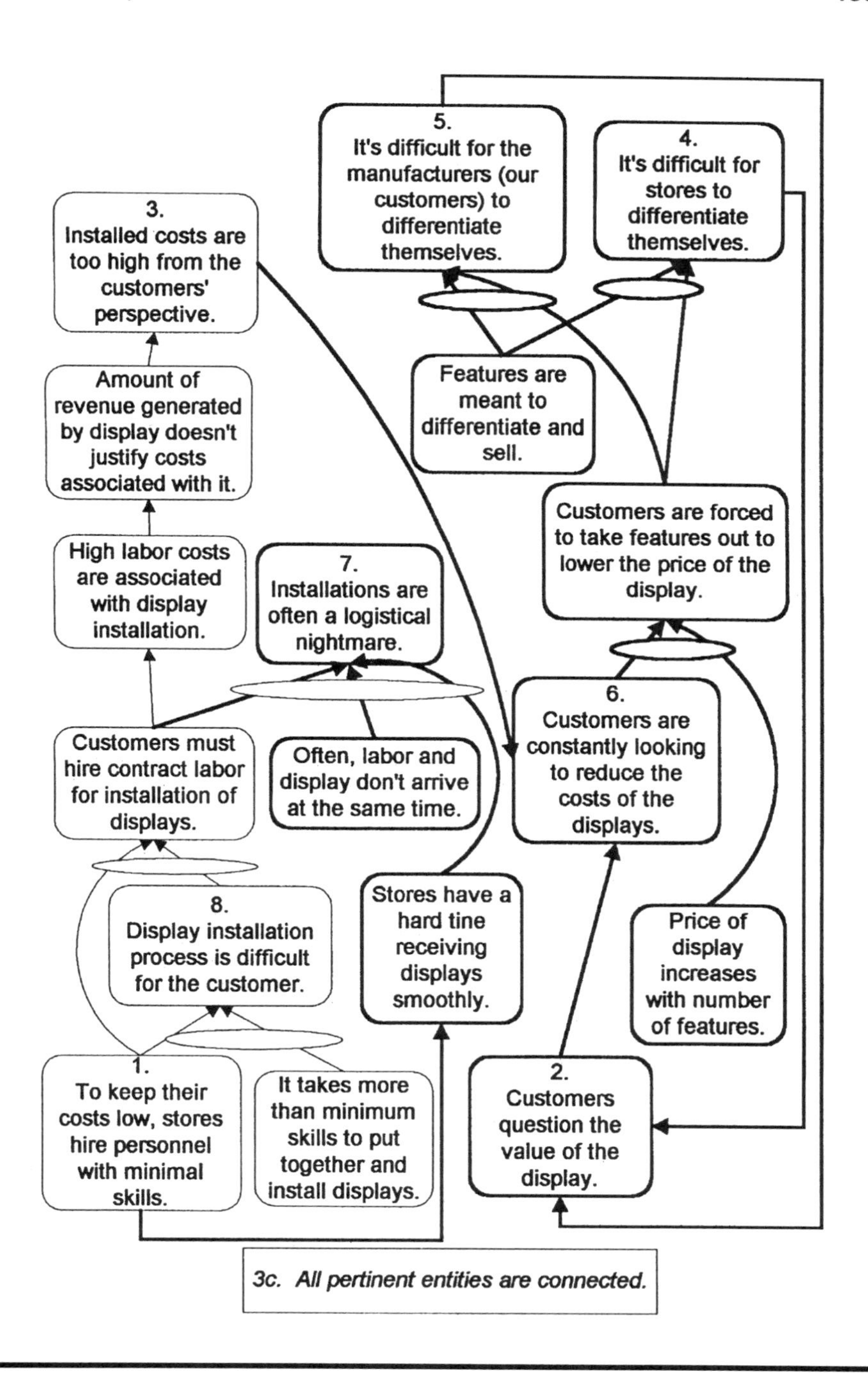

5.
It's difficult for the manufacturers (our customers) to differentiate themselves.
4.
It's difficult for stores to differentiate themselves.
3.
Installed costs are too high from the customers' perspective.
Features are meant to differentiate and sell.
Amount of revenue generated by display doesn't justify costs associated with it.
Customers are forced to take features out to lower the price of the display.
High labor costs are associated with display installation.
7.
Installations are often a logistical nightmare.
6.
Customers are constantly looking to reduce the costs of the displays.
Customers must hire contract labor for installation of displays.
Often, labor and display don't arrive at the same time.
Stores have a hard tine receiving displays smoothly.
8.
Display installation process is difficult for the customer.
Price of display increases with number of features.
1.
To keep their costs low, stores hire personnel with minimal skills.
It takes more than minimum skills to put together and install displays.
2.
Customers question the value of the display.
3c. All pertinent entities are connected.

Figure 8.7

fashion, immediately invoke the clarity reservation, and modify your tree so that it does.

1. Do a "gut reality check." Ask yourself if the tree, as a whole, represents a clear picture of the reality of the system. If yes, fantastic! If no, you have a reservation. Utilize the categories of legitimate reservation to uncover and diagram the "missing pieces." Sometimes, the "real" reality is different from the reality to which you and your team would rather admit. This step is asking you to do a gut check on whether the tree reflects the "real" reality, even if that reality is one which you have avoided.
   a. You might want to show your tree to a colleague.
      i. If you want to confirm that the tree is, in fact, a reflection of the real current reality, share it with someone who also "lives" that reality with you.
      ii. If you want a set of eyes to look at the tree from the perspective of uncovering assumptions that you may have missed, share the tree with a colleague who does *not* live the reality with you. Ask them to find the possible holes in your logic, listen to their reservations, and fill in the missing pieces that they will surely highlight for you.

Display Manufacturing (continued)

Display Manufacturing's tree looked like Figure 8.8 at this point in the process.

Step 5. Apply the "So What Test"

Your intuition led you to select an initial list of pertinent entities, and you used that list to develop the beginning framework of the tree. You then expanded your tree to make it a clearer picture of the system. Now it's time to look at the tree with a fresh set of eyes and really determine what's pertinent and what's not. You have spent a good deal of time thinking about these entities as they exist in relation to one another. This step will ask you to look at each entity as if it was related to nothing else except the system itself.

Consider your original list of pertinent entities as nothing more than entities that exist, along with other entities, on your tree. Now look at each entity in isolation and ask yourself the following questions. Highlight in some way (such as with colored marker) any entities for which your answer is "yes."

- If your tree is for the purpose of identifying a core problem, ask: *SO WHAT! If this entity wasn't caused by its purported cause(s), and if it wasn't causing the entities it was supposed to have caused, would I want to remove it from the system? Is it undesirable on its own merit? Is it undesirable simply because it exists?*
- If your tree is for the purpose of discovering a core strength, ask: *SO WHAT! If this entity wasn't caused by its purported cause(s), and if it wasn't causing the entities it was supposed to have caused, would I care about keeping it in the system? Is it desirable on its own merit? Is it desirable simply because it exists?*
- If your tree is for the purpose of discovering a general core driver, ask both of the above.

Don't be surprised to find yourself answering yes to only a few of the entities that were on your original pertinent entity list. Typically, the entities you select in this step include a few from your original list along with additional entities taken from your tree. When you are creating your first current reality tree, this step may be very difficult. Over time, I hope that you will agree with me that it is a most refreshing part of the process.

This is a critical part of the process, and I caution you against skipping it. The next step is identification of the core cause. This is where we truly determine "core cause of what?"

Display Manufacturing (continued)

Display Manufacturing put every one of the entities on its current reality tree through the "SO WHAT" test. The task was especially difficult, because they had to look at the entities not from the perspective of Display Manufacturing, but from the perspective of their customers. The revised list of pertinent entities became:

1. Display installation process is difficult for the customer.
2. Stores have a hard time receiving displays smoothly.
3. Installations are often a logistical nightmare.
4. Amount of revenue generated by display doesn't justify costs associated with it.
5. Installed costs are too high from the customers' perspective.
6. Customers question the value of the display.
7. It's difficult for the manufacturers (our customers) to differentiate themselves.
8. Displays aren't causing consumers to buy what they (displays) are meant to sell.

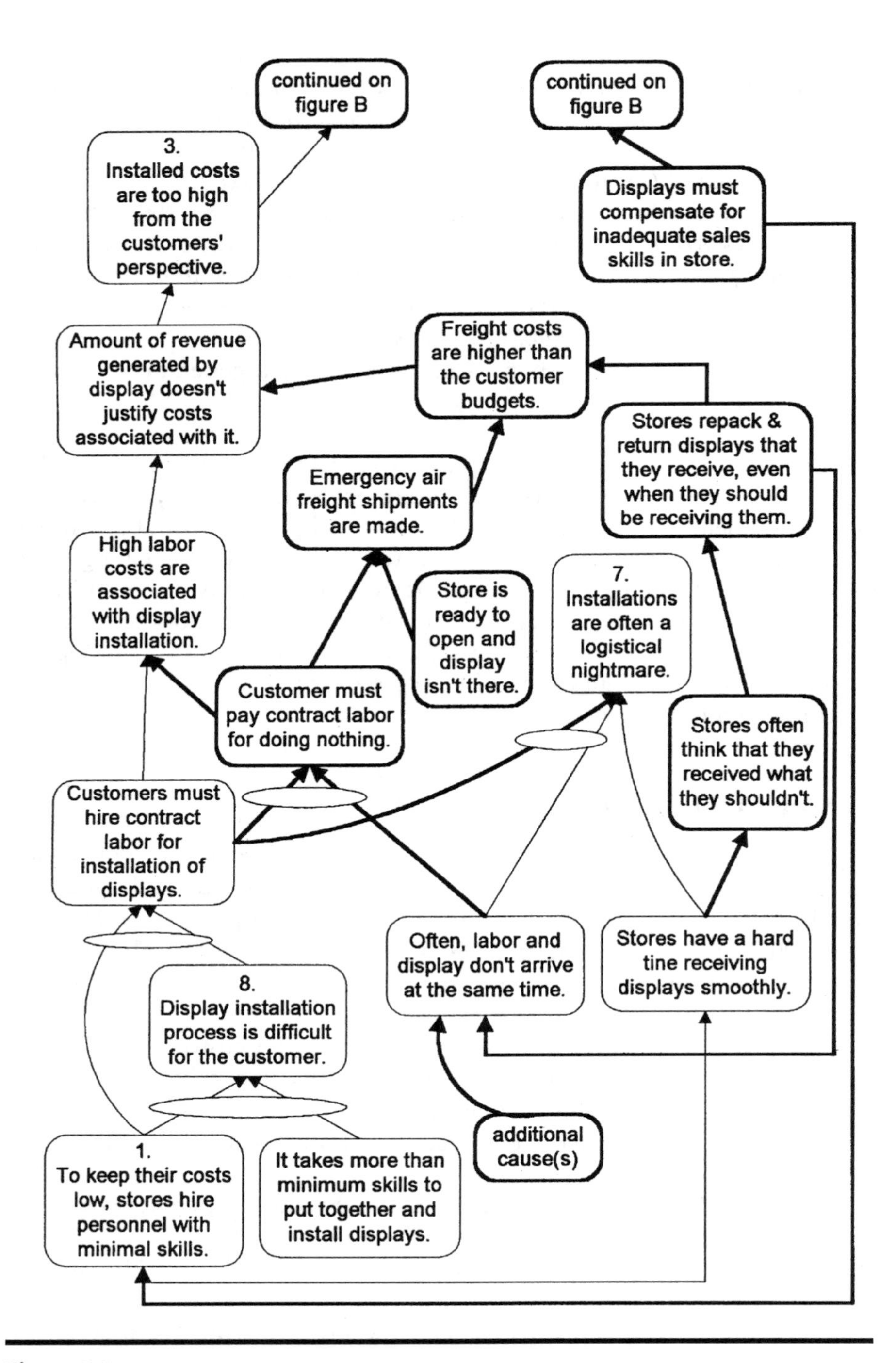
continued on figure B
continued on figure B
3. Installed costs are too high from the customers' perspective.
Displays must compensate for inadequate sales skills in store.
Amount of revenue generated by display doesn't justify costs associated with it.
Freight costs are higher than the customer budgets.
Stores repack & return displays that they receive, even when they should be receiving them.
Emergency air freight shipments are made.
High labor costs are associated with display installation.
Store is ready to open and display isn't there.
7. Installations are often a logistical nightmare.
Customer must pay contract labor for doing nothing.
Stores often think that they received what they shouldn't.
Customers must hire contract labor for installation of displays.
Often, labor and display don't arrive at the same time.
Stores have a hard tine receiving displays smoothly.
8. Display installation process is difficult for the customer.
additional cause(s)
1. To keep their costs low, stores hire personnel with minimal skills.
It takes more than minimum skills to put together and install displays.

Figure 8.8a

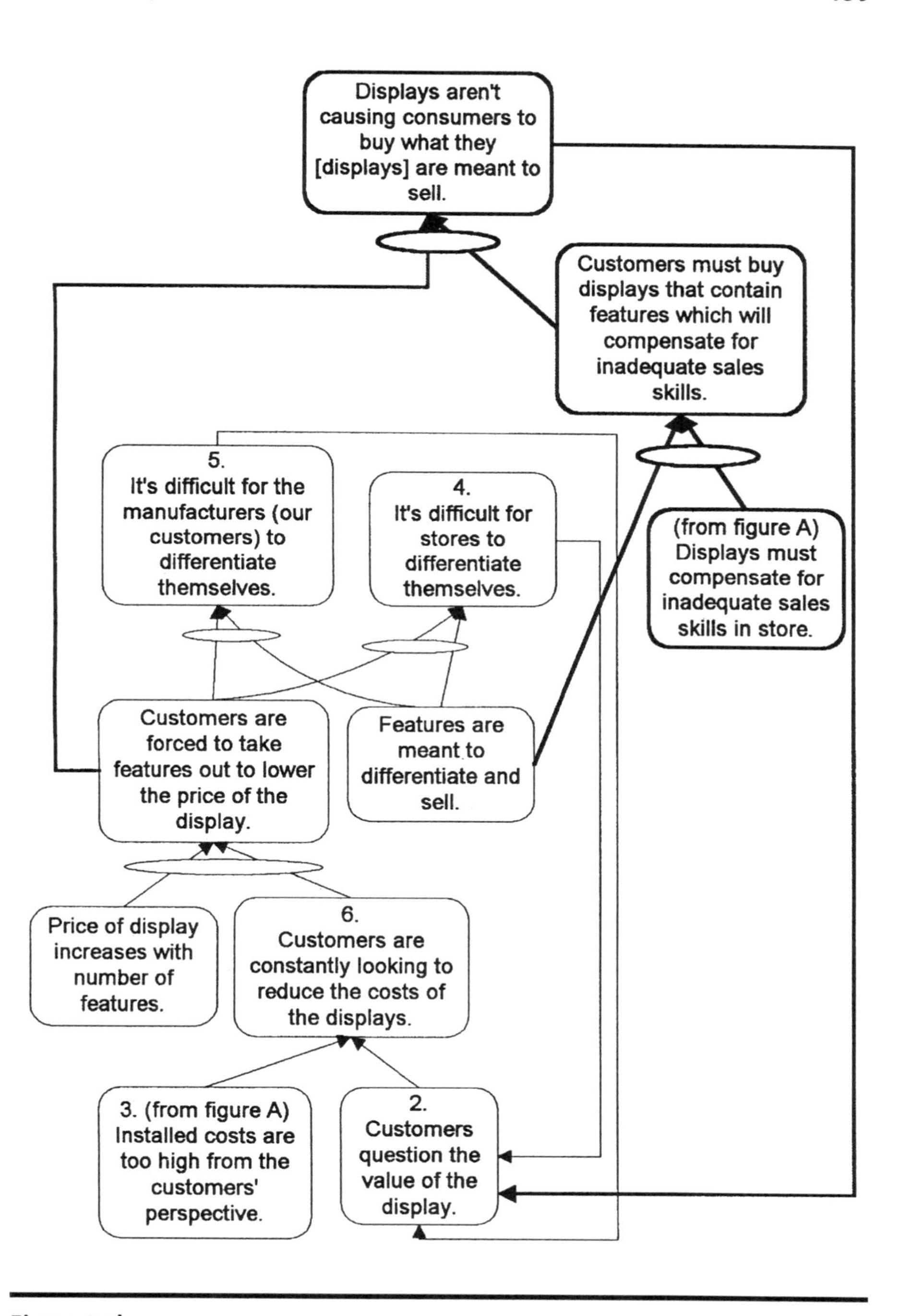
Displays aren't causing consumers to buy what they [displays] are meant to sell.
Customers must buy displays that contain features which will compensate for inadequate sales skills.
5. It's difficult for the manufacturers (our customers) to differentiate themselves.
4. It's difficult for stores to differentiate themselves.
(from figure A) Displays must compensate for inadequate sales skills in store.
Customers are forced to take features out to lower the price of the display.
Features are meant to differentiate and sell.
Price of display increases with number of features.
6. Customers are constantly looking to reduce the costs of the displays.
3. (from figure A) Installed costs are too high from the customers' perspective.
2. Customers question the value of the display.

Figure 8.8b

You may be surprised at some of the entities that are not on the above list. For instance, what about *Customers are constantly looking to reduce the costs of the displays?* Of course, Display Manufacturing didn't necessarily like this little fact of life, because it meant they were constantly justifying their prices. However, the question they needed to answer was, *would their customers perceive this as an undesirable effect?* Not likely. In fact, they'd probably be quite happy with their purchasing agents diligently looking after costs!

Step 6. Identify Core Cause(s)

Danny and I planted a tree in front of our house last year. Stems and leaves that almost look like mini trees grow from its base and roots. The landscapers call these mini trees "shooters." When the shooters are present, the base of the tree looks really messy, and it's difficult to differentiate between the trunk and the top of the tree. In fact, it almost looks like a bush! The shooters soak up water and other nutrients, which detract from the tree's overall rate of growth, in terms of both height and strength of trunk. Therefore, we trim these shooters when they appear. We're going to go through a similar activity in this step of the current reality tree process.

> The grand aim of all science is to cover the greatest number of empirical facts by logical deduction from the smallest number of hypotheses or axioms.
>
> Albert Einstein, 1950

A. First, we must distinguish the tree from the "shooters." We do this by highlighting the paths by which the reselected pertinent entities (the results of the "so what test") are connected to each other through linear and "V" connections, as illustrated in Figure 8.9.

We have now identified that the pertinent entities are connected with each other as indicated in Table 8.2.

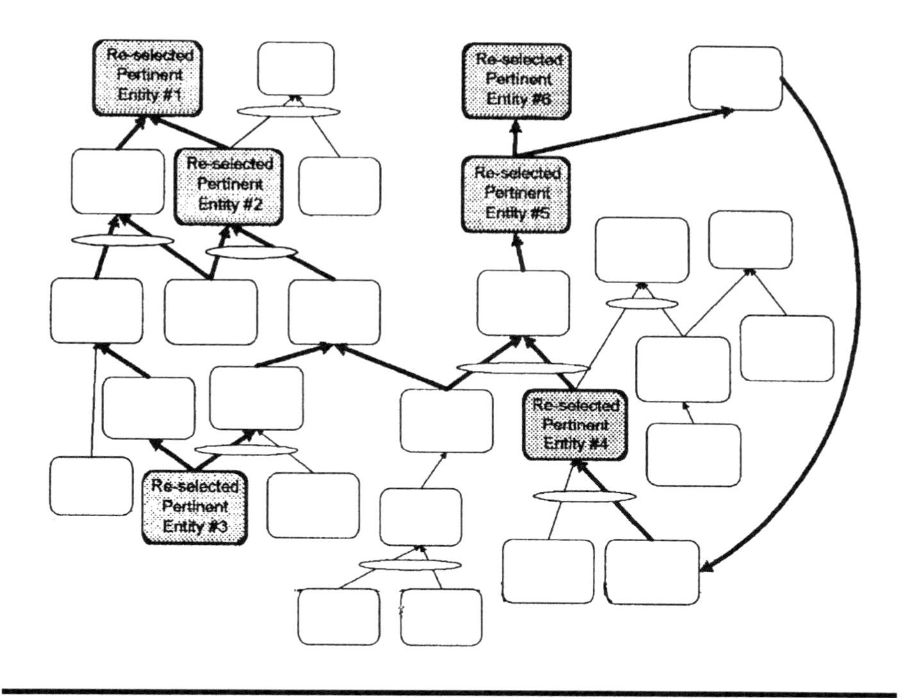

Figure 8.9

Table 8.2	
Entities Involved	***Type of Connection***
2,1	V — common cause is #3
2,1	Linear
2,1	V — common cause is entity directly below #2
5,6	Linear
4,5	Linear
3,1	Linear
3,2	Linear
2,5	V — common cause is entity to the left of #4
5,4	Linear
6,4	V — common cause is #5

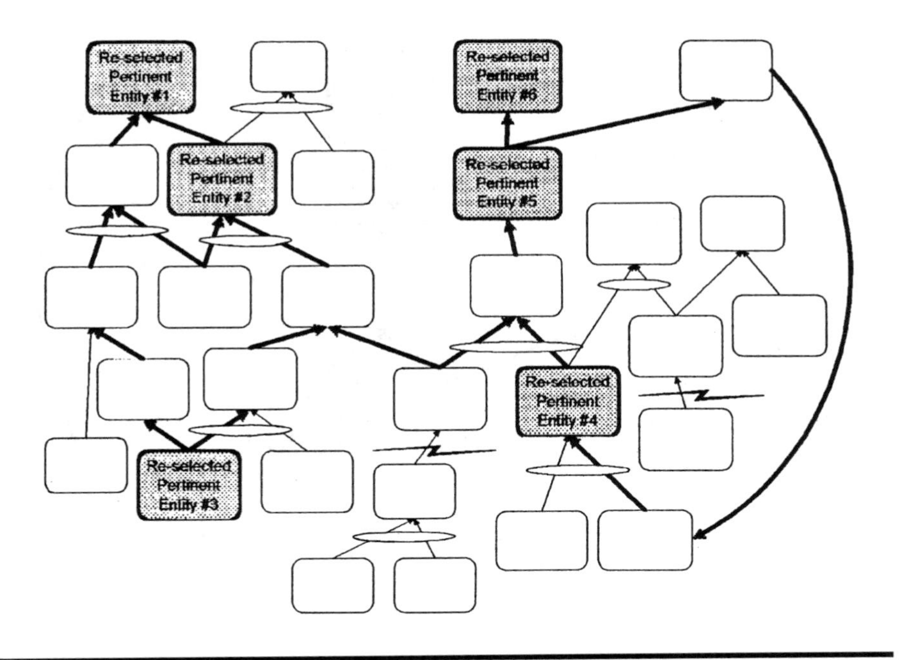

Figure 8.10

B. Trim entities that are not involved in connecting the pertinent entities with each other. In step 6A, we highlighted the silhouette of the current reality tree. What remains outside the silhouette are the "shooters" — those entities that, for the purpose of this analysis, cloud the picture. We trim these entities from the tree, just as Danny trims the shooters from the tree in our front yard.

   In Figure 8.10, the entities beneath the lightning bolts are to be trimmed from this tree.

C. For each entity that is now an entry point, determine the degree to which it is responsible for the existence of the pertinent entities. What percentage of the pertinent entities does the entry point cause? An entry point that is responsible for 80% or more of the pertinent entities is a core cause.

The same tree is illustrated in Figure 8.11, this time with the entry points labeled.

Now lets use Table 8.3 to look at the entry points and their relationship to the pertinent entities.

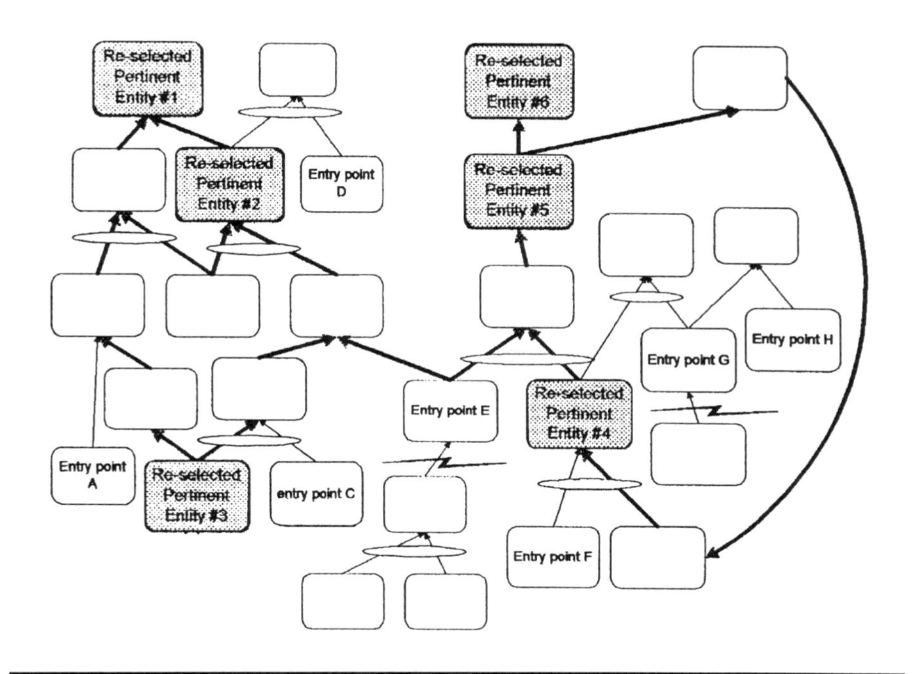

Figure 8.11

Table 8.3								
	1	*2*	*3*	*4*	*5*	*6*	*Total*	*% of 6*
A	X	—	—	—	—	—	1	16%
B	X	X	X	—	—	—	3	50%
C	X	X	—	—	—	—	2	33%
D	—	—	—	—	—	—	0	0%
E	**X**	**X**	**—**	**X**	**X**	**X**	**5**	**83%**
F	—	—	—	X	X	X	3	50%
G	—	—	—	—	—	—	0	0%
H	—	—	—	—	—	—	0	0%

D. Select the core cause. In this example, entity E should be selected as the core cause, as it is a cause for more than 80% of the pertinent entities.

- Double check it! Ask yourself, "If the core problem (in this case Entity E) were eliminated, what would happen to the rest of the tree?" I visualize a nonlinear game of dominoes, as I imagine the effects of the core cause vanishing, and their effects vanishing as well.

At this final point in the current reality tree process, you may have some of the following questions:

What if there is no entity on the tree that is responsible for more than 80% of the pertinent entities? This is the only time you should concentrate on "diving down." Look for a common cause for one or more of the entry points to your tree. Let's imagine that entity E only caused 75% rather than 83% of the pertinent entities. I would then look for a common cause of entity E and, perhaps, entity 3 (entry point B), because such a cause would be a common cause for all 6 of the entities. Another place to look would be for a common cause of entry points E and F, or F and 3 (entry point B). In fact, the small cluster of entities that we previously trimmed would likely be of real assistance in a case like this.

I have built this current reality tree to identify a core problem. What if I have identified something that I really don't believe I can solve? If you identify a core problem that you really want to solve, but that you believe you are not empowered to solve, please take a look at the other thinking process tools. The evaporating cloud is designed to help you find breakthrough, simple, practical solutions — I'll bet there's one lurking out there that you haven't thought of yet. The future reality tree is designed to help you think through a complete solution. The starting point for many future reality trees is an idea generated with the evaporating cloud. The prerequisite tree is designed to help you determine how to overcome obstacles to putting in place a desired solution — the more difficult the task is to implement, the more useful the prerequisite tree is! **The bottom line is, if you want to fix it, you can, and there are tools to lead you there right in this book!**

I have built this current reality tree to identify a core problem. What if I have identified something that I simply don't want to solve? First, distinguish between *don't want* and *can't*. If this is a "can't," then see my advice above. If this is a case where you simply don't want to touch that entity, check to see if there are any additional entities that meet the 80% criteria. If so, you might choose to select one of them. If not, seek one by using the method for "diving down," which is described above.

The core cause that I identified is no surprise to me. In fact, if I were your teenager, I'd probably say something like, "No duh!" Did I waste my time doing this tree when I already knew the answer? Many times, the core cause is something you have already suspected. It might even be one of the entities on your initial pertinent entity list. There is the possibility that the answer to this question might go back to the reason you decided to build a current reality tree in the first place. If you started out already believing strongly that you *knew* the answer to the questions that the tree is meant to provide, then I would agree that you wasted your time. On the other hand, ask yourself the following: Do you have a clearer understanding of the system now than before you started the tree? Are you now in a better position to communicate the realities of the system to others? If you have built the current reality tree to identify a core problem, have you been working or focusing the efforts of your resources at solving the core problem (the weakest paradigm or policy link in the chain)? Or are resources scattered about fighting the fires that result from its existence? If you have built the tree to identify a core strength or competence, has the system really been utilizing it in beneficial ways? Is the system wasting it or exposing it to competitive deterioration?

One more time: If you already believe strongly that you know the answer to the question(s) that the tool is designed to answer, do not waste your time picking up the tool!

Sometimes you will end up confirming what you intuitively knew before you picked up the tool, and sometimes you will be surprised with the core cause.

Display Manufacturing (continued)

Let's take a look at Display Manufacturing's current reality tree as it evolved through this last step in the process (Figure 8.12). The heavy lines show the paths by which the reselected pertinent entities are connected to each other. There were only two "shooters" to be trimmed. The entry points are noted with wide arrows, and the reselected pertinent entities are highlighted with thick lines.

It was obvious to the team that the core problem they were dealing with was "*To keep their costs low, stores hire personnel with minimal skills.*" But this was also something they didn't want to address directly. Who were they to try to change their customers' customers' hiring practices? They decided then to take a look at the entity the core problem combined with as it led to pertinent entity #1: *It takes more than minimum skills to put together and install displays*. They also asked themselves the "cause insufficiency" question on the connection between the core problem and pertinent entity #2. This led to their verbalization that only when the minimal skills are combined with the existing entity, *It takes more than minimum skills to smoothly receive displays*, do the stores have a difficult

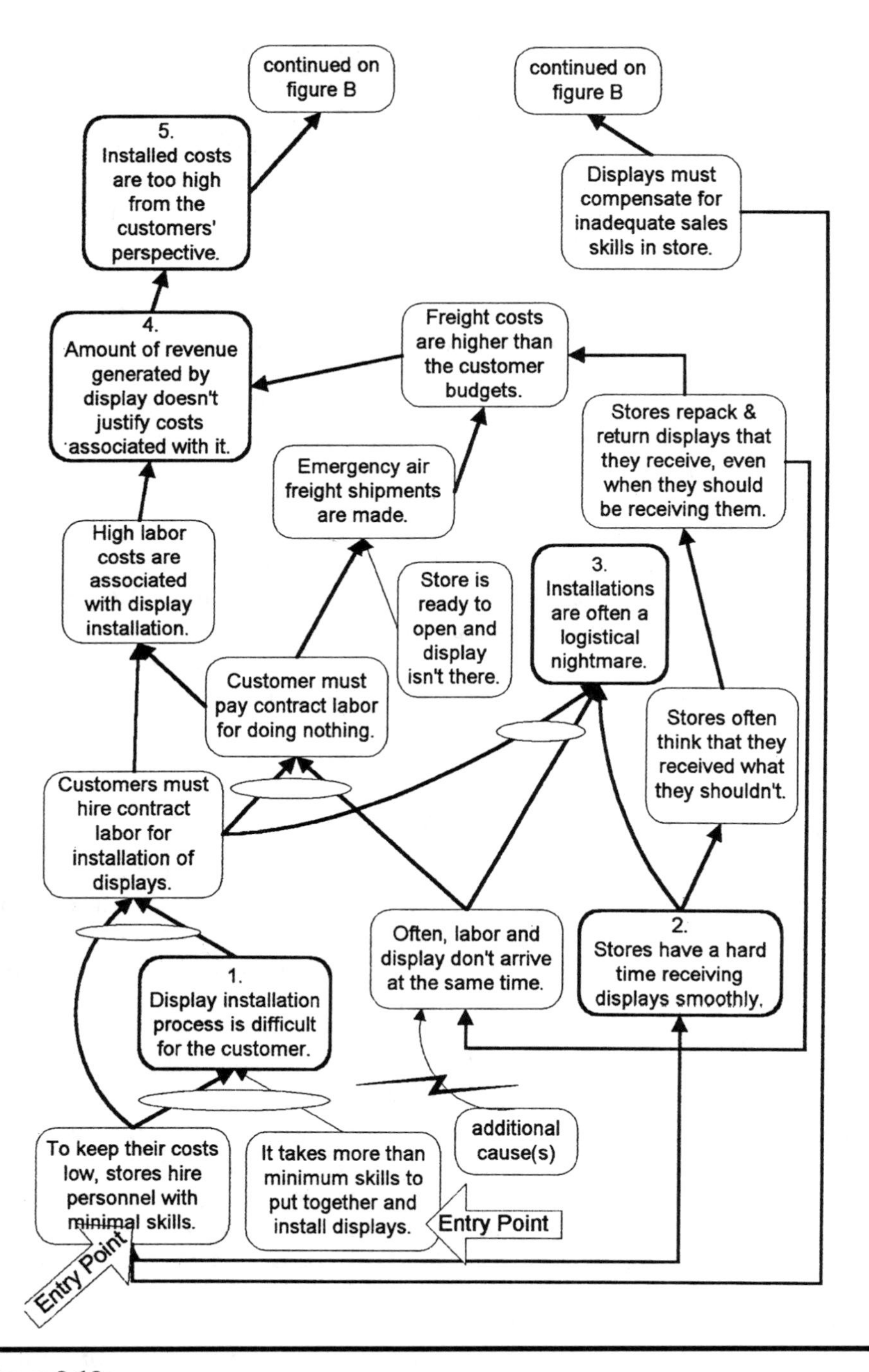

continued on figure B
continued on figure B
5. Installed costs are too high from the customers' perspective.
Displays must compensate for inadequate sales skills in store.
Freight costs are higher than the customer budgets.
4. Amount of revenue generated by display doesn't justify costs associated with it.
Stores repack & return displays that they receive, even when they should be receiving them.
Emergency air freight shipments are made.
High labor costs are associated with display installation.
Store is ready to open and display isn't there.
3. Installations are often a logistical nightmare.
Customer must pay contract labor for doing nothing.
Stores often think that they received what they shouldn't.
Customers must hire contract labor for installation of displays.
Often, labor and display don't arrive at the same time.
2. Stores have a hard time receiving displays smoothly.
1. Display installation process is difficult for the customer.
additional cause(s)
To keep their costs low, stores hire personnel with minimal skills.
It takes more than minimum skills to put together and install displays.
Entry Point
Entry Point

Figure 8.12a

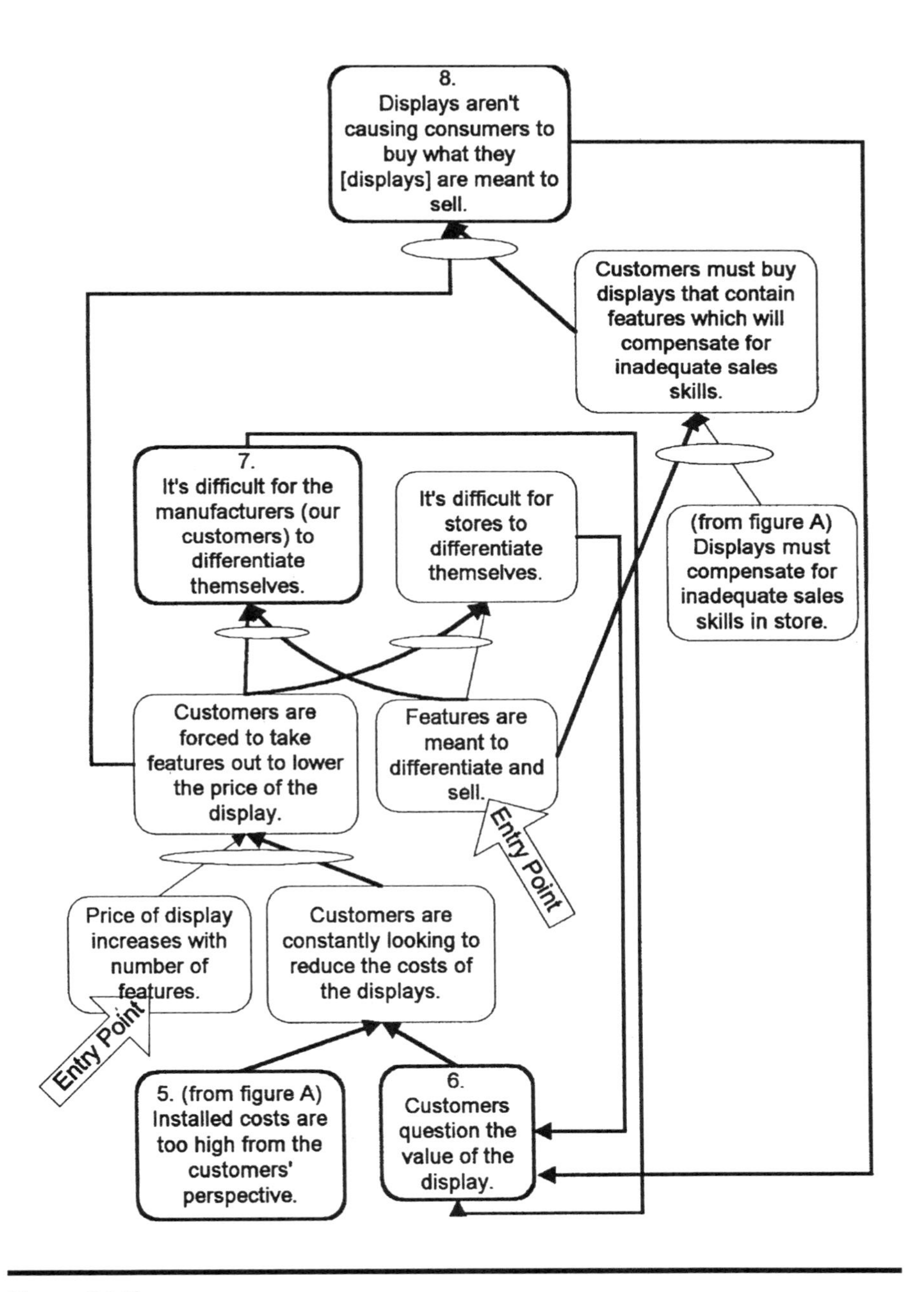
8.
Displays aren't causing consumers to buy what they [displays] are meant to sell.
Customers must buy displays that contain features which will compensate for inadequate sales skills.
7.
It's difficult for the manufacturers (our customers) to differentiate themselves.
It's difficult for stores to differentiate themselves.
(from figure A) Displays must compensate for inadequate sales skills in store.
Customers are forced to take features out to lower the price of the display.
Features are meant to differentiate and sell.
Entry Point
Price of display increases with number of features.
Customers are constantly looking to reduce the costs of the displays.
Entry Point
5. (from figure A) Installed costs are too high from the customers' perspective.
6.
Customers question the value of the display.

Figure 8.12b

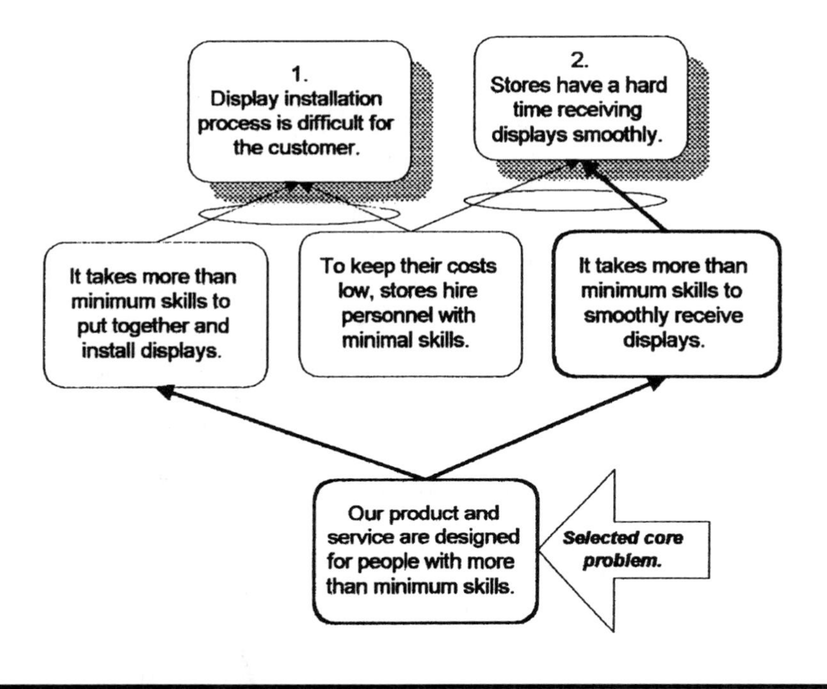

Figure 8.13

time receiving. Realizing that the nature of *their* product design and service was a common cause to both of those, and that this new common cause was a core problem they could get very excited about solving, they quickly moved into solution development mode and derived a unique, simple, yet high-value solution. (If you're interested in that solution, see their future reality tree in Chapter 7. The revised bottom portion of the current reality tree is shown in Figure 8.13.

Next Steps

Now that you have completed your current reality tree, ***do something with it!*** If you have used the tool to identify a core problem, it's time to solve that problem. If you are confident that you know a workable solution and equally confident that you know just what to do in order to implement that solution, just do it! If not, you will find other thinking process application tools designed to help you through these issues.

- If you have no idea what the solution might be, if the solutions that come to mind have been tried before (and you face the reality

that in spite of those past attempts, the core problem still exists), or if your ideas seem to violate other system policies or paradigms, then select the evaporating cloud as your next step.
- Perhaps you have an idea for the solution to the core problem but sense that it needs some thinking through. Maybe you suspect that there are potential undesirable consequences, or that although your idea is a good start, it won't be sufficient in and of itself to result in the elimination of the core problem and the creation of an environment where the pertinent undesirable entities are replaced with pertinent desirable entities. In such cases, the future reality tree will guide you through building a full, robust solution.
- If you are confident that you have a solution defined, but you don't know how in the world you are going to implement it — you just don't see the path from today's reality to that future reality that contains your solution — reach for the prerequisite tree.
- Finally, if you know what to do, but have a feeling that your specific action plan can use a little bit of thinking through, the transition tree will guide you through developing a step-by-step action plan.

You may have developed your current reality tree simply for the pleasure of learning and understanding your environment a little more. If so, good for you! I will still challenge you to do something with that learning. *Learning doesn't happen until behavior has changed.** Ask yourself in what ways this learning has and will change you and how you act in and outside of the system. Ask yourself who else might benefit from the learning, and communicate what you've learned with them. (Note, you may want to do a transition tree prior to communicating sensitive issues with others.)

* A saying of Bob Pike, Creative Training Techniques, Minneapolis, Minnesota.

Chapter 9

Evaporating Clouds

When we deliberate it is about means and not ends.

Aristotle, 4th Century B.C.

The evaporating cloud is by far the most often used of the thinking processes. This may be due to the fact that it is the easiest of the tools to learn. The cloud only has five entities, and it takes just a few minutes from start to finish, once you've learned the technique. The tool is used for conflict resolution, and one thing we humans are not short on is conflict.

I know that as soon as I say, "Every problem that exists can be described as a conflict, in the form of an evaporating cloud," somebody will bring one to my attention that cannot. So I will say it this way: I have not found a problem yet that cannot be described as a conflict, in the form of an evaporating cloud. Further, I have not found a problem yet that is impossible to solve. (I have, unfortunately, found problems that people have been unwilling to solve.)

The cloud views a conflict as a set of five relationships (see Figure 9.1).

1. Arrow #1 identifies entity B as a perceived necessary condition for entity A.
2. Arrow #2 also identifies entity C as a perceived necessary condition for entity A.
3. Arrow #3 identifies entity D as a perceived necessary condition for entity B.

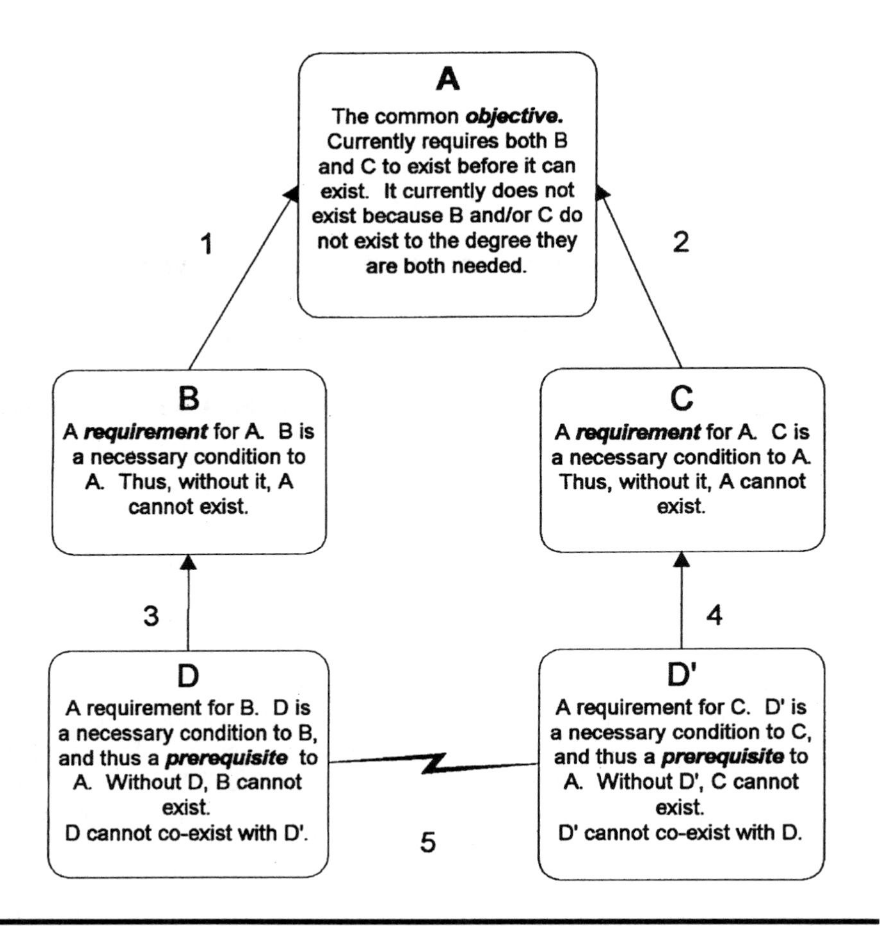

Figure 9.1

4. Arrow #4 identifies entity D′ as a perceived necessary condition for C. (Please note that the notation D′ is often interchanged with the notation E.)
5. Arrow #5 identifies entities D and D′ as entities that are believed to be in conflict, perceived as unable to coexist in the current system.

Once a way is found to invalidate just one of these relationships, the conflict is dissolved, and the cloud is evaporated.

The Process

The major steps of the evaporating cloud process are:

1. Articulate the problem and diagram the cloud.
2. For each arrow, uncover assumptions and identify potential solutions, using the necessary condition thinking process described in Chapter 5.
3. Choose an injection to implement.

As we go through each of the steps in detail, I encourage you to work on a real problem — one in which you are currently involved and would like to solve.

1. Articulate the problem.
   a. The first thing you must do is recognize that you are involved in a problem (or conflict, or dilemma), it is a problem you really want to solve, and you have not yet found a satisfactory solution for it. *If you don't want to solve the problem, don't bother picking up the tool.* Make the decision to take a few moments to step back and observe the problem for what it is — a conflict that contains at least one assumption that is invalid. Once you have made the commitment to solve this problem, you have empowered yourself to examine all sides of the issue, including those assumptions that you may have held near and dear to your heart. You might call this taking ownership.
      - Tell yourself the story of the problem. You may find that sometimes it is helpful to do this in writing. A short paragraph will suffice. This helps you start to get things into perspective. There is no need to take more than a moment or two here. If you find yourself taking more time, you are *over* analyzing before you even get started. If you continue along this path, you'll do a good job of getting yourself quite frustrated. Stop immediately and go to the next step.
      - You may have the tendency to try to solve a problem before you have given yourself the opportunity to define it with the diagram. *Solving the problem comes later.* First, you are giving yourself the opportunity to look at the problem itself. If you find that the act of verbalizing the problem automatically leads your mind to generate some solutions, write the ideas that come to mind on a separate sheet of paper. This way, you won't forget them, and it will be easier to continue with the process. Then, remind yourself that solving it comes a few steps down the road and get back to the process.
   b. Diagram the problem as the five entities of an evaporating cloud. The cloud and each of its entities are illustrated in Figures 9.1 and 9.2. *The order in which you fill in the boxes doesn't matter.*

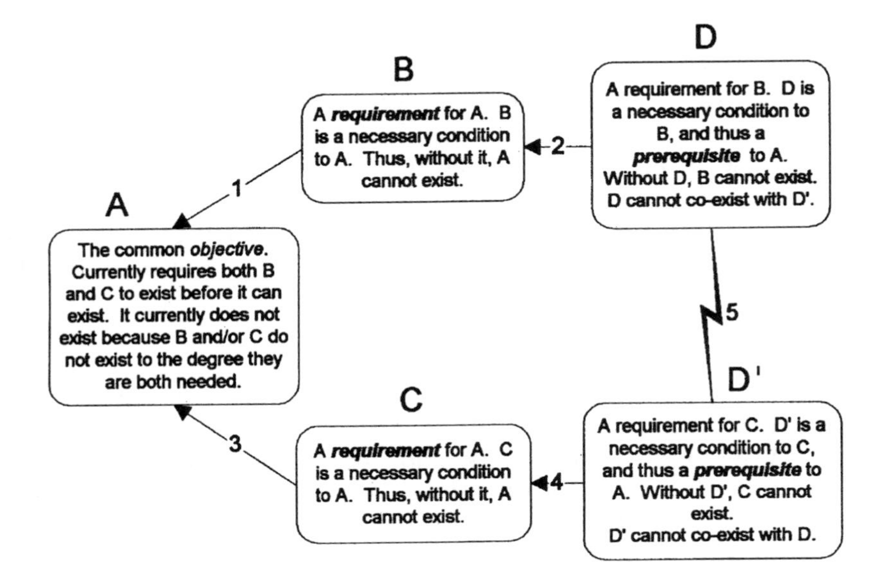

Another common way that the evaporating cloud is diagrammed.

Figure 9.2

It only matters that all five boxes are completed, and that when you read the diagram, it is a clear and concise articulation of the problem. (Later in this chapter, I describe a common method for constructing the cloud.)

- You should be able to read a cloud using the language of necessary conditions, such as:
  - In order to have [objective entity A], [requirement entity B] must exist.
  - We also need [requirement entity C] in order to have [objective entity A].
  - We can't get [requirement entity B] until or unless we have [prerequisite entity D].
  - We must have [prerequisite entity D′] in order to get [requirement entity C].
  - [Entity D] and [Entity D′] are mutually exclusive. They cannot coexist. (In order for [entity D] to exist, [entity D′]

must not. In order for [entity D′] to exist, [entity D] must not.)

2. For each arrow, uncover assumptions and identify potential solutions using the necessary condition thinking process described in Chapter 5.
   - In step one, you built a model of the problem that should, when you read it back to yourself as necessary condition relationships, articulate the problem. In this step, you will methodically search each arrow and identify the assumptions that lurk beneath each of those arrows. These assumptions are the reasons the conflict exists and are your keys for solving it in a win–win–win fashion.

   a. For arrows one, two, three, and four, use the process described in Chapter 5 to surface the assumptions that form the necessary condition relationships.
   b. For arrow five, surface the assumptions by answering any of the following:
      - Why can't entities D and D′ co-exist?
      - Why aren't we allowed to have both D and D′?
      - Is there *any* overlap between the two?
      - Are they *really* mutually exclusive? Why?
      - You should be able to read the conflict relationship back to yourself in any of the following ways:
        - The reason we are unable to have both D and D′ is [assumption].
        - D and D′ are not allowed to coexist in this environment because [assumption].
        - It must only be D *or* D′ because [assumption].
   c. For each assumption that you have listed in (a) and (b) above, ask yourself the following two questions:
      - Is the assumption that is currently being made in the conflict's system (situation)? If the answer is no, scratch it off your list. If the answer is yes, move to the next question.
      - Is this assumption *valid* in the conflict's system (situation)? If the answer is no, you have evaporated the cloud! You have found the paradigm constraint of the situation and a viable path out of the conflict!

3. Brainstorm injections, using the processes described in Chapter 5.
   a. For every assumption that remains on your list as a result of step 2d, brainstorm injections. Remember, the operative word here is *brainstorm*. At this point in the process, avoid judging the ideas for practicality. Simply list as many ideas as you can think of. The more ideas the merrier. I try to make the time to

first write down all the ideas I can think of, and then push myself to write down at least two more. Even if I need to let it sit for a day or two before I can come up with those last two ideas. That's when the simple, practical, second-order solutions seem to happen for me. After I've emptied my head of all the things I've already thought of, consciously or unconsciously, my thinking path is clear enough to allow new ideas to emerge.

Nothing is more dangerous than an idea, when it is the only idea we have.

Alain, 1908

---

b. You can test injections aimed at the assumptions of arrows one through four by filling in the blanks to either of the following statements:
   - If I/we decide to implement [injection], then we will be able to achieve [objective of necessary condition relationship] without the [necessary condition].
   - If I/we decide to implement [injection], then we will not need [necessary condition] in order to achieve [objective of the necessary condition relationship].

c. You can test injections aimed at the assumptions of arrow five by filling in the blanks to either of the following statements:
   - If I/we decide to implement [injection], then there will no longer be a conflict between [entity D] and [entity D′].
   - If I/we decide to implement [injection], then [entity D] and [entity D′] will be able to coexist.
   - If I/we decide to implement [injection], then we will no longer need [entity D] and/or [entity D′].

4. Choose an injection to implement.
   a. Now that you've done all of this thinking, it's time to select one of your injections and solve your problem in reality! The injection you select should meet *all* of the following conditions:
      - You really want to see it implemented.
      - You are willing to put forth the effort to make it happen.
      - Select simple and practical, instead of difficult and complex.

b. Your injection may be simple enough to "just do it." Chances are, though, that once you make your selection, you might want to do some planning. The Thinking Processes offer additional tools, depending on what your next step should be.
   - If you think you should explore the idea further to ensure that what you ultimately implement is robust enough to achieve the objective *and* avoid undesirable consequences, then utilize the Future Reality Tree (Chapter 7).
   - If you perceive difficult obstacles that must be overcome in order to get the injection implemented, then utilize the Prerequisite Tree (Chapter 10).
   - If you just want to put a little more thought behind how to implement the injection, then utilize the Transition Tree (Chapter 6).

The Evaporating Cloud in Action: Three Real Life Case Studies

The following stories are real cases that illustrate the use of the evaporating cloud in three very different situations.

The Cloud of the Cloud

The name *evaporating cloud* has been controversial. You may be asking, "Where did such a name come from?" *Jonathan Livingston Seagull*, the best seller by Richard Bach, was a book that had a profound impact on Dr. Goldratt. Bach later authored *Illusions: The Adventures of a Reluctant Messiah*. In *Illusions*, the main characters remove storm clouds from the sky by thinking them away.

> If you really want to remove a cloud from your life, you do not make a big production out of it, you just relax and remove it from your thinking. That's all there is to it.*
>
> Richard Bach, 1977

---

* Bach, Richard, *Illusions: The Adventures of a Reluctant Messiah*, Dell Publishing, 1977.

Imagine a problem as a giant storm cloud. Now, imagine that you can evaporate that problem into thin air simply by thinking about it. As a physicist, Dr. Goldratt believes there is no such thing as conflict in nature, so any conflict we experience is our own doing. His teaching acknowledges that every problem is a conflict, and that conflicts arise because we create them by believing at least one erroneous assumption. Thus, simply by thinking about the assumptions that enforce the existence of a conflict, we should be able to resolve any conflict by evaporating it with the power of our thinking. Goldratt decided to name the conflict resolution tool *Evaporating Cloud* in honor of Richard Bach.

Many consultants and academics who are teaching and using the thinking processes have changed the name of this tool from evaporating cloud to things like conflict resolution diagram, conflict diagram, and dilemma tree. Their reasons for doing this seem sound enough: *evaporating cloud* isn't really a professional or scientific sounding name; it doesn't describe its function as well as the names of the other thinking process tools, such as current reality tree or future reality tree. This is a serious tool, and the name *evaporating cloud* is, well, too "fluffy" to be taken seriously; especially by *executives*! Until last spring, I was in this camp and was referring to the tool as the dilemma tree.

Dr. Goldratt was a keynote speaker at a conference I had helped to organize. One evening during the conference, he joined the conference committee for dinner. The subject of the evaporating cloud became the topic of conversation. Dr. Goldratt was clearly angry and offended about the variety of names that have surfaced for this tool. I chalked up his anger to ego and joined several people around the table who attempted to reason with him about the merits of the other names. Dr. Goldratt then reiterated the history of the name. He went on to explain that in his culture, when something is named in honor or in memory of a person, the only person in the world with the right to change that name is the person in whose honor the thing was named. Thus, even if he agreed with the rationale for the name change, Dr. Goldratt himself doesn't have the right to change the name of the evaporating cloud. From his perspective, the only person with that right is Richard Bach, and anybody who calls it anything other than evaporating cloud is committing an act of disrespect to both Dr. Goldratt and Mr. Bach.

This left me in a personal predicament. I was one of the consultants who felt somewhat unprofessional talking about poofy clouds with senior executives. I did believe that the name *evaporating cloud* was an obstacle to its acceptance as a viable tool. For several years, I had been referring to the evaporating cloud as either the conflict diagram or the dilemma tree. I was in the process of writing this book, and I was planning to

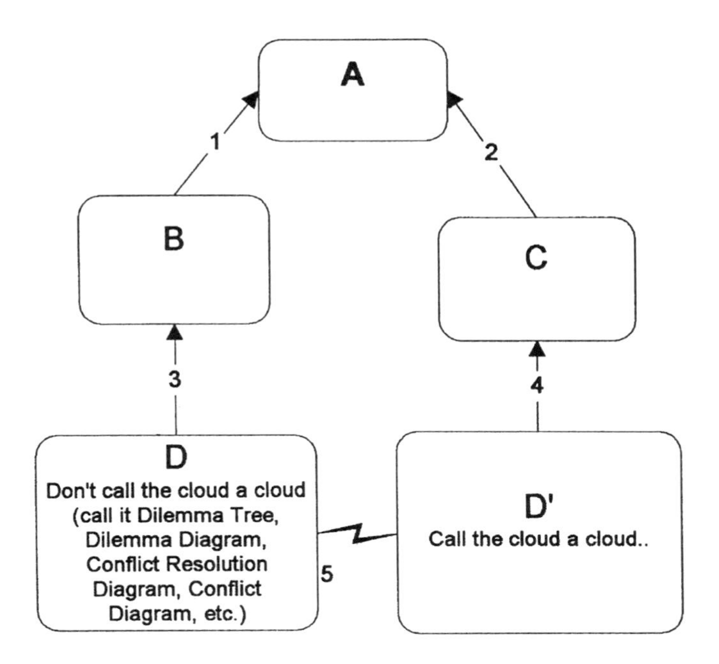

Figure 9.3

introduce the term *Dilemma Tree* to the world. Yet, the last thing in the world I wanted to do was personally offend my friend and mentor Eli Goldratt. So, I pulled a pen from my purse, a clean napkin from the table, and drew "the cloud."

Figure 9.3 shows the five-entity structure of the cloud. I began with the boxes labeled D and D′. These are the parts of the cloud that describe the elements of the situation that are perceived to be in conflict. I asked myself, "So, Lisa, what's the conflict?" The answer to that was easy enough — it was to call the tool a cloud or not, and I articulated this as illustrated.

The next question I asked myself was, "What need does calling the cloud a cloud allow us to satisfy?" I wrote my answer for this question in the "C" box. Then, for the "B" box, I had to answer, "What need is satisfied by calling the cloud something else (Figure 9.4)?

Finally, I had to clarify the common purpose. If there is no common purpose, there is no conflict. What ties all of this together? What are all of the parties in this situation attempting to accomplish, that require both respect for Dr. Goldratt and all of the tools having names descriptive of their processes? It was obvious to me that the common objective was to

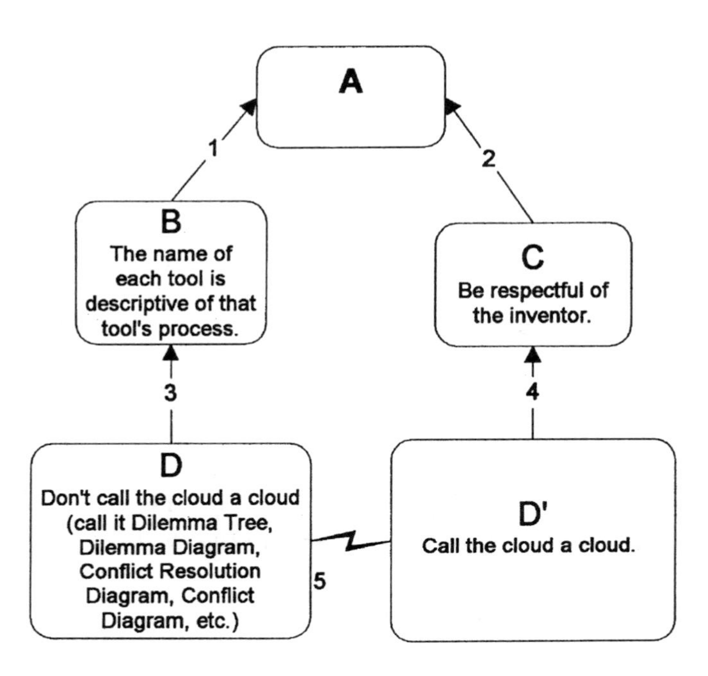

Figure 9.4

successfully spread the principles and practices of TOC. I wrote my answer as Entity A.

The cloud was complete. Now came the real work. How to evaporate it?

If I could find a way for effective spread of TOC without requiring all of the tools to have names that describe their respective processes, *or* without the need to be respectful of Dr. Goldratt, there would be no conflict. If I could find a way that allowed all of the tools to have names that describe their processes without requiring a name change for the cloud, there would be no conflict. If I could find a way to be respectful of Dr. Goldratt without the necessity to continue to call the cloud a cloud, there would be no conflict. If I could find a way to have it both ways — to change the name *and* not change the name — there would be no conflict.

My next task was to identify the assumptions that formulated each of the five relationships. If I were to identify an assumption that was already invalid, then, just like in *Illusions*, the cloud would evaporate just because I thought about it. If I were unable to identify an already invalid assumption, then I would need to pick one, and decide what to do in order to make it invalid in the future.

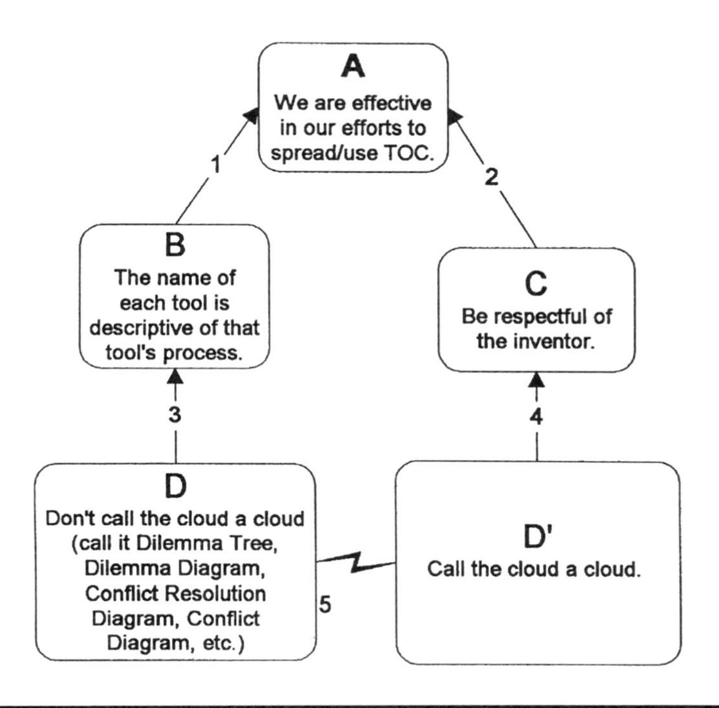

Figure 9.5

> One's belief that one is sincere is not so dangerous as one's conviction that one is right. We all feel we are right; but we felt the same way twenty years ago and today we know we weren't always right.
>
> Igor Stravinsky, 1959

Here is the list of assumptions that I came up with. In the hope that I would find a way to convince Dr. Goldratt to change *his* mind about the subject, I began by exploring the relationships on the right side of the diagram.

Arrow #2: We must be respectful of the inventor in order for us all to be effective in our efforts to spread/use TOC because:

- *He is deserving of such respect.*
- *I care about his feelings.*

- *His name is linked strongly to the TOC philosophy and applications.*
- *My own personal code of conduct prohibits disrespect and views conscious disrespect as unprofessional and unethical.*

Not seeing any assumptions that were obviously erroneous, or any for which a simple and practical solution called to me, I moved on to the next relationship.

Arrow #4: The reasons we must call the cloud a cloud in order to be respectful of the inventor are:

- *From Dr. Goldratt's perspective, Richard Bach is the only person in the world with the right to change the name, and to do so without his permission constitutes tremendous disrespect toward both Dr. Goldratt and Mr. Bach.*
- *It's not respect if he doesn't perceive it that way.*

Again, I saw both of these assumptions as valid in this situation. There was no mistaking Dr. Goldratt's position. One idea that came to mind was to contact Richard Bach, explain the situation to him, and ask him to request Dr. Goldratt to change the name of the tool. Before I seized that as THE solution, though, I pressed on. If I could find an assumption that was already invalid, I wouldn't have to go through the trouble.

Arrow #5: We can't call it a cloud *and* call it other things because:

- *The market will be confused by more than one name.*

This assumption had been validated for me many times, and it appeared that we TOC consultants and professors were already doing a good job of confusing the market. When teaching the thinking processes to anybody who has read Dr. Goldratt's book, *It's Not Luck**, the question inevitably came up, "so when am I going to learn this evaporating cloud thing?" This happened even if they had already been taught the tool under the name *dilemma tree!* In fact, in most of my thinking process courses, I made sure the students knew that the dilemma tree was, in fact, the same thing as an evaporating cloud (and all the other names that I am aware of) — just in case they ever communicated with someone who had read *It's Not Luck* or had been taught the thinking processes by someone other than me.

I then remembered Kathy Suerken. Kathy is the president of the nonprofit organization, TOC For Education, Inc. The mission of Kathy's

---

* Goldratt, Eliyahu M., *It's Not Luck,* North River Press, 1994.

company is to help educational systems improve through using TOC, and her intent is to spread the TOC principles and tools far and wide throughout educational systems all over the world. In just a few short years, Kathy has proven that she's got a very good chance at success. Many school districts in and outside of the States are using the thinking processes to reengineer their administrations, and teachers are using them to design their curricula. One of the thinking process application tools — yes, the evaporating cloud — is spreading like wildfire and is rapidly becoming the tool of choice for many peer-mediation programs. These are programs in which students help each other resolve their conflicts. The kids call it "clouding." Just maybe, the parents of one or more of these kids will read this book. It would be nice if they both had the same name for the same tool. I certainly don't want to give any teenagers reason to argue with their parents over something like the name of the tool they're using to resolve arguments!

Having not found an erroneous assumption on "Dr. Goldratt's" side of the cloud, and clarifying some real reasons to continue to call it a cloud, my own dilemma intensified. This meant that it was time for me to examine the side that, in my mind, I had been defending for three years. I started at the bottom.

Arrow #3: We must call the cloud something other than a cloud in order for each tool to be descriptive of that tool's process because:

- *The name "evaporating cloud" doesn't describe what the tool does.*

At the time I was trying to solve my dilemma, I did believe this assumption to be true, and pressed on. Later, upon reflection, I have come to appreciate that the name "evaporating cloud" *is* quite descriptive of the tool's function. In fact, a recent student, a project manager for an information technology company, told me that *evaporating cloud* describes what the tool does very well. I have since asked many people — from shop floor personnel to senior executives — and the vast majority have absolutely no problem with the name. Gulp.

Arrow #1: It's impossible to be effective in our efforts to spread/use TOC unless the names of each tool describe that tool's process because:

- *The tool is otherwise difficult for us to describe or for the prospect to grasp.*
- *A "cutesy" name presents major obstacles to selling or teaching the tool, particularly to executives.*
- *We are uncomfortable using a "fluff" term, such as evaporating cloud, in our roles as consultants and educators.*

Uh oh. As I wrote these down on the napkin, I felt my face begin to flush. I realized that it wasn't Dr. Goldratt's ego that was enforcing this particular conflict, it was my own, and that of the rest of us TOC consultants and academics out there! In reality, the evaporating cloud is by far the easiest of the thinking process application tools to teach or learn. It can be explained in just a couple of minutes, taught in a matter of hours. The name has never been an obstacle for someone who is looking to invest in learning an effective conflict resolution tool — though it certainly seems to have been an obstacle from the perspective of the "sellers." So the question I faced was, would it be easier for me to swallow my ego and face reality, or should I call Richard Bach and see if he would put forth some sort of public request to change the name of the tool that was named in his honor?

I decided to face my own erroneous assumptions and embrace once again the name *evaporating cloud*. I also hope that this book will encourage other consultants and teachers to do the same.

The Case of the Equipment Manufacturer: Purchasing vs. Manufacturing vs. Sales

A manufacturer of temperature-control equipment for the injection molding industry was plagued by parts shortages in manufacturing. Manufacturing was blaming Purchasing for poor vendor management. Purchasing was blaming Manufacturing and Sales for taking too long to provide the information they needed to purchase products within their vendors' quoted lead times; and, of course, Purchasing and Manufacturing were under the gun to keep inventories as low as possible. The company was poised for tremendous growth, but unless it got the parts situation under control, it was going to have a tough time competitively fulfilling additional demand.

Figure 9.6 illustrates their cloud. Would you have worded it the same way? Maybe yes, maybe no. It's OK either way — as long as what you've written is a clear articulation of the problem.

We surfaced the following assumptions, which were the reasons for the persistent conflict:

- Arrow #1: In order to have the right materials when we need them, we must wait until we have all of the correct information *because we manufacture to order.*
- Arrow #1: In order to have the right materials when we need them, we must wait until we have all of the correct information *because it is too risky, financially, to stock components.*

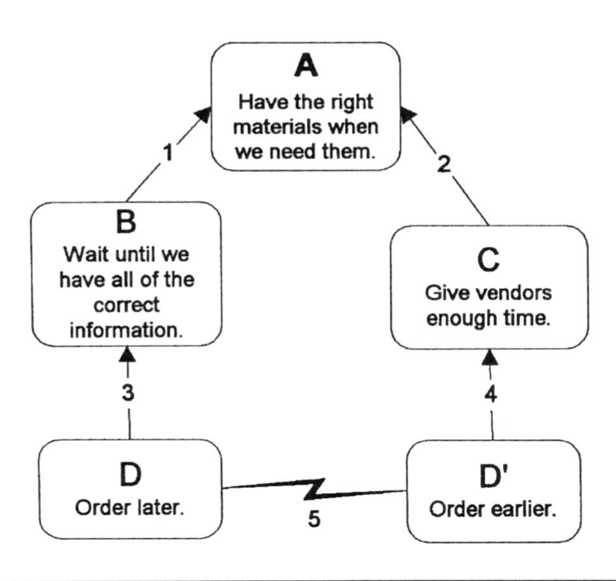

Figure 9.6

- Arrow #2: In order to have the right materials when we need them, we must give vendors enough lead time *because the vendors' lead time is longer than zero.*
- Arrow #3: In order to wait until we have all the correct information, we must order later, *because it takes a long time for engineering and/or customers to determine all the details of the product's final design.*
- Arrow #4: In order to give vendors enough time, we must order earlier, *because the lead time our customers want is longer than the lead times our vendors offer for purchased components.*
- Arrow #5: It is impossible to order later *and* order earlier, *because all parts carry equal weight in cost and risk.*

Three of the assumptions jumped out at us as erroneous. Although this company's products are custom goods from their customers' perspective, they are assembled to order. The vast majority of components are standard and fairly low cost. This meant that holding a little bit more inventory in the stockroom (and a lot less inventory in work in process) would pose little risk for the company and would result in the ability to drastically shorten lead time for customer orders. The assumptions under arrow number one were, therefore, invalid. When it became clear that arrow one was invalid, we took a look at arrow five. Once the company had the inventory replenishment policies in place for the standard components, purchasing would have the capacity to respond more effectively to the few specialty items. Direction could also be provided to sales and engineering on the crucial design elements that might require longer lead times, so that when possible, those elements could be addressed earlier.

By looking at the issues in this way, they were able to come up with a simple, practical solution. They were able to identify their paradigm constraints that held firm the policies that each side tried to enforce, which just kept the fuel of the conflict raging. They were also able to identify what *they* could do in order to be in a more competitive situation, rather than trying to implement a solution that would require changing their vendors or their customers.

The Towel Bar Battle

My daughters, Jennifer and Rachel share a bathroom. When you enter their bathroom and turn right, you face the tub. Between the entrance and the tub, on your right, is a wall. On the wall are two towel bars, side by side. Jennifer (then 15 years old) and Rachel (then 9 years old) argued, almost daily, and sometimes more, over who had the rights to the towel bar closest to the tub. Rachel argued that she needed her towel to be closest to the tub, because without it, she got the floor all wet. Jennifer yelled at Rachel when she left the floor wet after her bath.

Jennifer argued that she needed *her* towel to be closest to the tub, because without it, she used up precious time in the morning and risked being late for school. (I know, I know — I never said teenagers were logical.) One day, the girls brought their argument to me in an attempt to get me to take sides. I told them I was not going to solve it for them, but if they wanted to find a solution, I would be happy to teach them a way to do so.

Me: *Jennifer, would you like to learn how to really solve this?*
Jennifer: *No!* (exit Jennifer to her room that we refer to as "Jenn's cave." Door slams, and loud music, if you can call it that, is now heard through her wall.)
Me: *Rachel, would you like to learn how to really solve this?*
Rachel: *Yes, Mom.*
Me: *OK, Rachel, what is it that you want?*
Rachel: *What do you mean, Mom?*
Me: *Well, Rachel, what are you guys fighting about? What did the two of you come in here hollering about?*

Rachel: *Who gets the towel bar.*
Me: *OK. So tell me, what do you want?*
Rachel: *I want the towel bar next to the bathtub!*
Me: *Great. Write that down, just like you said it, in the box with a "D" in it. Now, what does Jenn want?*
Rachel: *You know, Mom. Jenn wants the same thing I want!* ***She*** *wants the towel bar next to the bathtub, too!*
Me: *OK, Rach. Write that down in the box with a D′ in it.*
Me: *OK, Rach, why is it so important for you to have the towel bar that's right next to the tub?*
Rachel: *Because if I don't, then I drip all over the place when I get out of the tub, the floor gets all wet, and Jennifer yells at me!*
Me: *I want to make sure I understand you right, honey. The reason you feel you need the towel bar next to the tub is so that you won't get the floor all wet?*
Rachel: *Yes, Mom.*
Me: *OK, write that down in the "B" box. "I won't get the floor all wet."* (Rachel writes it.) *Now, why does Jennifer need the towel bar next to the tub for her towel?*
Rachel: *Because she doesn't want me to have it. She just wants to be mean.*
Me: *Sorry, Rach, I don't think Jennifer would agree with you that she's doing this just to be mean. Let's not try to read her mind, OK? What is she saying? What does she tell you that her reason is?*
Rachel: *So that she's not late for school.*
Me: *Well, if that's what she's saying, that's what you need to write in the C box, Rachel.*

So far, Rachel's cloud looks like Figure 9.7.

Me: *OK, Rach, you're almost finished. Just one more box to go.* (I cover the D and D′ entities with one hand.) *Why is it so important that you don't get the floor all wet and that Jenn is on time for school?*
Rachel: *Well, so Jenn won't yell at me, Mom.*
Me: *Rach, do you think that's what both of you want out of this situation?*
Rachel: *Yes, Mom, I do.*

I then asked Rachel to write that in the A box, which she did. Her completed cloud looked like Figure 9.8.

I reviewed the cloud with her. *OK, so both of you want Jenn to not yell at you. In order for that to happen, you've got to keep the bathroom floor dry, and Jenn has to be able to be on time for school. In order for*

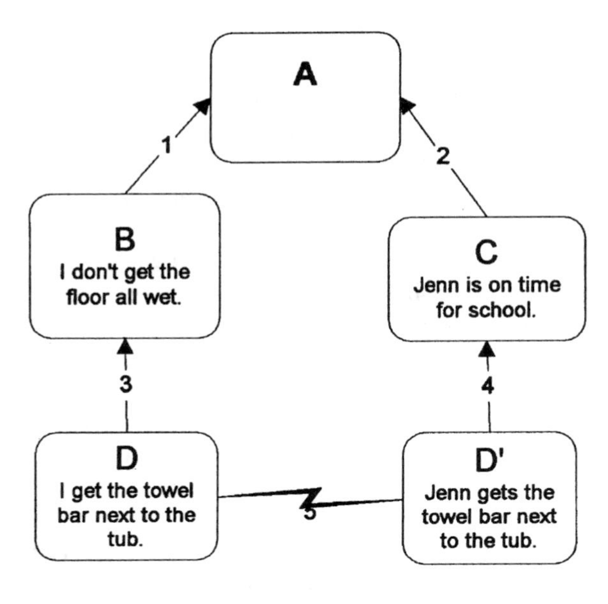

Figure 9.7

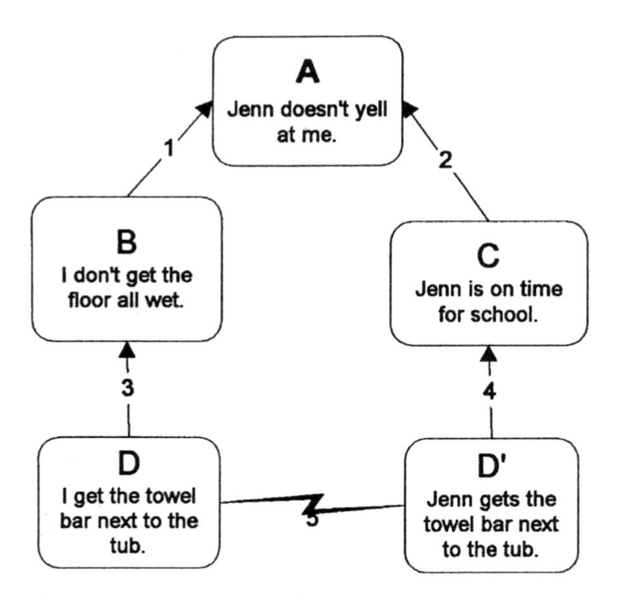

Figure 9.8

you to avoid getting the floor all wet, you believe that you need to have your towel on the bar next to the tub, and in order for Jenn to be on time for school, she wants to have her towel on the bar next to the tub. (I point to the B and C entities, and hide the D and D′ entities.) *Well, isn't this interesting. You don't want to get the floor wet, and Jennifer wants to be on time for school. These two things surely don't seem to be a problem.* (I remove my hand from the D and D′ entities.) *But look, Rach. It's only when we see what the two of you, individually, are trying to do about each of them, that the trouble begins!*

I went on to explain to Rachel that she could solve the problem by figuring out a way to break only one of the arrows. *Look, Rach, if you can figure out a way to keep Jenn from yelling at you even if you do get the floor all wet, the problem goes away. If you are able to keep the floor from getting wet without having the towel bar next to the tub, the problem goes away, too. Right?* Rachel nodded. I think she was waiting for me to reiterate some of my past suggestions about how to solve the problem, like laying her towel next to the tub before she started her bath, but, I didn't. I really wanted her to come up with her own solution, so I continued my explanation. My plan was that after I finished my explanation, I would leave her with the piece of paper to brainstorm her own injections, and then get back together with her. I went on with similar explanations for arrows two and three, and then explained breaking the conflict arrow. *Now, if you can figure out a way for both of you to have the towel bar next to the tub, or for neither of you to have the towel bar next to the tub....*

At that moment, Rachel interrupted me. *Mom! Mom! I've got it! How about if you just get another towel bar and hang it under the one that's already next to the tub! Then we'd* ***both*** *have our towels next to the tub!* After a little bit of discussion, we agreed that if Jennifer also like the solution, then I would go out and buy dual towel bars for their bathroom.

Rachel was excited that she had solved the problem. The next step was to share her solution with her big sister. Rachel asked me to do this with her. We knocked on Jennifer's door. *What do you want!"* came the voice from "the cave." *Your sister has solved your problem. Would you like to hear her solution?*

Jenn: *No! I know that I'm going to have to take the other towel bar! I always have to be the one to give in because I'm the oldest! Not this time!*

Me: *Well, Jenn, this time you're lucky, because Rachel is the one who solved it and not me. Her solution has nothing to do with you giving in to anything. Would you like to hear it?*

Jenn: *Fine.*

Rachel then showed Jennifer the cloud that she built and explained her solution. After Jennifer confirmed that I would really buy new towel bars, she reluctantly agreed that Rachel's solution was, indeed, win–win. They haven't argued about towels again.

The girls still try to bring me many of their arguments, but more and more often, I remember to send them off to do the cloud. It works.

A Common Method for Constructing the Cloud

The D and D′ entities are the hot buttons of the conflict. They are what is fought over in the situation, and they are probably the focus of attention. Because of this, it is usually easy to verbalize the D and D′ entities of the cloud. This method has us begin to construct the cloud with those entities.

1. In the "D" box, write the answer to the question, "What does one side want?"
2. In the D′ box, write the answer to the question, "What does the other side want?"
   - The D and D′ entities should clearly be opposites or entities that are clearly mutually exclusive.
   - A word of caution: Don't get hung up on your initial wording of the entities. Now is not the time to be politically correct, or even grammatically correct. You are taking a first pass at articulating a problem, and it really is OK to erase and rewrite as you go along. My experience is that when people worry about the wording of entities on the first touch of pencil to paper (or fingers to keyboard), the process takes much longer than necessary. We are so conditioned to *"get it right the first time"* that we feel as if we've failed somehow if we find the need to make changes. Allow yourself to use not only a pencil, but an eraser, too!

The next steps define the reasons why each side wants what it wants. Why are the entities D and D′ required?

3. In the B box, write your answer to the questions "Why does the side that wants D want it?" Or, "What is it that won't happen unless D happens?" Or, "What will D allow that side to achieve?"
4. In the C box, write your answer to the questions, "Why does the side that wants D′ want it?" Or, "What is it that won't happen unless D′ happens?" Or, "What will D′ allow that side to achieve?

Now it's time to fill in the A box. This entity represents the common objective for B and C. *If there were not a common objective, there would be no conflict.* This objective is currently not fully attained, because the system does not contain all of B and all of C. Likely, the current solutions or policies are some sort of compromise. A little bit for you, a little bit for me, ensures that neither of us gets what we want or need.

5. In the A box, write your answer to the question, "What is the common objective for both B and C?" Or, "What is it that we can't achieve without B and C?" This might be a difficult entity for you to articulate. There are a few reasons for this. One is that sometimes the common objective is a more global objective than the two parties are conditioned to consider. The other is that we tend to make things more complicated or difficult than they need to be. Thus, you might find yourself tossing out the first, or most obvious, answer to the question, and struggling needlessly with "what must be the right answer."

Congratulations! You have just done a first draft of a cloud. Read the cloud back to yourself, and modify your wording so it reads back as described in step 1.3.1. Does the cloud reflect the conflict accurately? Once it does, congratulations again! You have built an evaporating cloud! Don't forget to evaporate it!

Chapter 10

Prerequisite Tree

> Tomorrow lurks in us, the latency to be all that was not achieved before.
>
> Loren Eiseley, 1978

On May 25, 1961, President John F. Kennedy established a goal for the United States that is still considered to be one of the loftiest ever: that the United States would establish itself as the leader in space science and technology by landing a man on the moon and returning him safely to earth, "before this decade is out." He cautioned the country that "if we are to go only half way, or reduce our sights in the face of difficulty... it would be better to not go at all."

While Congress and the public were, in general, enthusiastic about such an incredible undertaking and supported it with the necessary funds, NASA engineers and scientists realized the magnitude of the task. There were many obstacles — so much to be discovered, understood, and tested in order to go from where they were (knowing how to orbit the earth) to where they were now tasked to go (landing on the moon and coming back in one piece).

The goal was accomplished July 20, 1969. The Eagle landed, and Neil Armstrong took the first steps on the lunar surface, claiming, "That's one small step for man, one giant leap for mankind!"

While most of us won't be tasked with figuring out how to make the next greatest leap in science or technology, we are often assigned projects or dream of accomplishing things that seem far out of reach or just too difficult to achieve. In these circumstances, most of us do one of three things:

1. We lessen the goal to something that seems more within our grasp, compromising what we *really* want,
2. We decide to "forget about it," and keep it in our pile of never-to-be-achieved dreams, or
3. We continue on as we always have, under the delusion that somehow, someway, if it is meant to be, it will happen. "Someday, my dream will come true."

There are a few determined people in the world who actually don't compromise. They set their goals, determine what milestones they need to accomplish along the way, and make them happen. Every once in a while, they regroup, reassess where they are relative to achieving their lofty goal, adjust the milestones, and leap forward once again. The goal is always in mind.

The prerequisite tree is a tool that can make that elite circle of people much larger. The process of the prerequisite tree leads us to define what's in our way — obstacles — and what we need to make happen so nothing keeps us from achieving our goals. Some applications of the prerequisite tree include:

- Project planning
- Implementation planning
- Personal development plans
- Business process development and definition
- Marketing strategies
- Organizational strategies

A simple prerequisite tree is illustrated in Figure 10.1.

The prerequisite tree is a diagram that describes the necessary condition relationships that are involved in achieving objectives (see Chapter 5). When you create a prerequisite tree, you will utilize necessary condition thinking to describe the path or paths that must be taken in order to accomplish those defined objectives, or goals. The entities in the diagram all fall under the rules for necessary conditions, and are referred to as:

Objective. The objectives are entities that describe the goals of the prerequisite tree. These are what the system is going to accomplish as a result of attaining all the entities in the tree.

Intermediate Objective. The intermediate objectives are entities that describe milestones that must be accomplished in order for the objectives

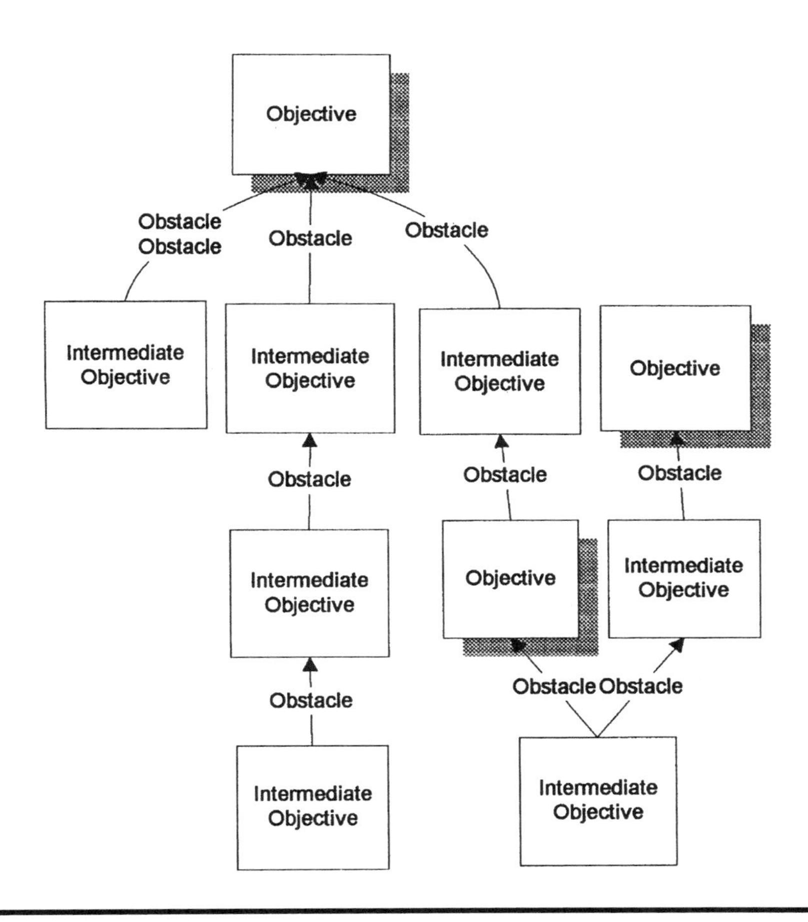

Figure 10.1

to be accomplished. Each intermediate objective is created in order to overcome an *obstacle* that stands in the way of achieving the objective/s, thus becoming a necessary condition to achieving another intermediate objective or objective.

Obstacle. Each arrow identifies a necessary condition relationship between an intermediate objective and objectives, or between an objective and another objective. This indicates that the objective or intermediate objective at the base of an arrow *must* be in existence before the objective or intermediate objective at the point of the arrow will be allowed to exist. The assumption behind the dependency is the *obstacle*, which is stated right on the arrow. Unless the obstacle is overcome somehow, the system will be unable to achieve the stated objective/s. Obstacles are stated as entities that exist in the current reality.

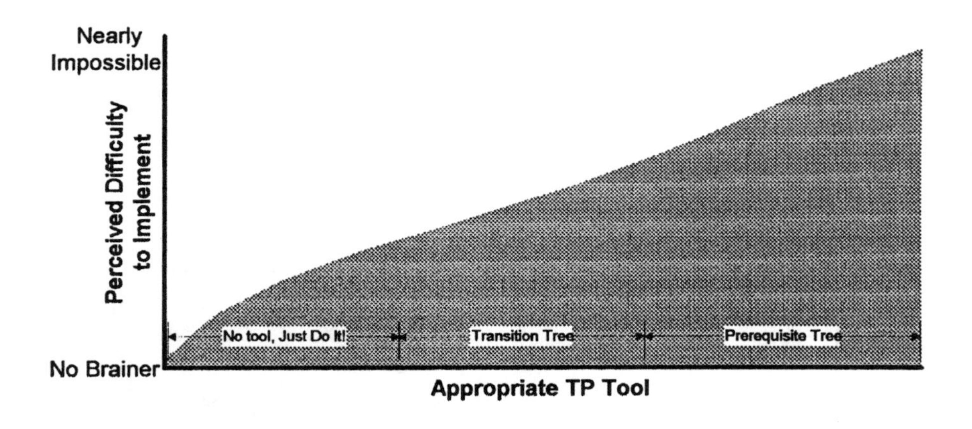

Figure 10.2

If you already have a pretty good feel for what steps you should take and the first things coming to your mind are *actions*, the prerequisite tree might be overkill. The prerequisite tree is the tool to use when you need to create an action or implementation plan, when it seems that all you are faced with are the reasons that your objective will be difficult to accomplish (Figure 10.2).

The Process

The major steps of the prerequisite tree process are:

1. Define the purpose for the prerequisite tree.
2. List the obstacles to achieving each of the objectives, and the intermediate objectives that will overcome them.
3. Map the implementation order of the intermediate objectives.
4. Implement!

Do you have any goals that you have put off because you thought they were just too hard to achieve? Pick one, and do the tree while we go through the steps that follow.

First, here are two general guidelines.

- **Verbalize entities in present tense terms.** Stating the entities as if they exist in the present accomplishes a few things. It helps you to project yourself mentally into the future and better visualize what you're thinking about. It will also keep you grounded in the realities of the real obstacles that exist, rather than those you perceive would exist given a certain course of action (that you haven't yet decided to take!). Additionally, when most of the entities on a tree contain the word "will" (as in *we will be more profitable*), the tree tends to be more cluttered and confusing for both the tree's creator and its readers.

- **Do not use tentative types of phrasing such as *maybe, might,* or *possibly*** (as in *we might not obtain the funding*) in the prerequisite tree. As soon as you feel yourself reaching for the M-word ("maybe"), recognize that you have an opportunity to decide what to do to create the conditions you seek, given the existing conditions. In the prerequisite tree, the M-word is an indicator that you have additional obstacles to surface, and thus additional intermediate objectives.

1. **Define the purpose of the prerequisite tree.** In this step, you will define why you are creating the prerequisite tree and establish its overall objectives.
   a. **Establish the premise for the prerequisite tree.** What are you planning? Take a moment to write down (or at least say to yourself) what you are about to plan. Is it a strategic plan, a new business process, or your own personal development plan? Are you building a new plant, planning your retirement, or implementing a new way of working in your organization? This helps you to focus on the task at hand. I find statements such as the following to be quite helpful:
      - *This prerequisite tree will define our organization's strategic plan.*
      - *This prerequisite tree will determine what we need to accomplish, and in what order, in order to implement the injections from our future reality tree.*
      - *This prerequisite tree will define my account plan for ABC Corporation.*
      - *This prerequisite tree will flush out what needs to happen in order for us to have our new product manufactured easily in the plant, and selling very well in the market.*
      - *This prerequisite tree will define the major things that need to happen in order for us to close the books at the end of each month within two days.*
      - *This prerequisite tree will determine what we need to accomplish in order to get to the moon.*

b. **Define the objective/s for the prerequisite tree.** Once you have made the decision to create a plan, do what Stephen Covey says, and *Begin with the end in mind.** What is the purpose of the plan? What is it that you expect to accomplish? How will you know when the implementation or the project is complete and successful? President Kennedy's objective was for the United States to be the leader in space. Write the objectives of the transition tree as entities, using present tense terminology. Here are a few examples:
   - I am retired at the age of 50.
   - We have a new plant in Rio.
   - On-time shipping performance is consistently 99 to 100%.
   - My book is published.
   - Our new product is in distribution in Europe.
   - We have 12 new engineers.
   - ABC Corporation's ERP system is providing their management with the information needed to make sound decisions.
   - One or more injections from a future reality tree should be objectives of a prerequisite tree when they meet the criteria described earlier in this chapter (difficult to implement). Additionally, a prerequisite tree is appropriate when your future reality tree contains several injections, and it is important to determine the order in which they should be implemented. This is an issue when you don't have unlimited resources with which to carry out the implementation of the injections.

   Avoid suboptimization when selecting your objectives. This means you should look at the big picture, and take other stakeholders into consideration. Will owners, employees, colleagues, family, customers, vendors, or your community suffer from undesirable consequences should you achieve the objective? If so, reconsider it! (See Chapter 7.)

2. **List the obstacles to achieving each of the objectives and the intermediate objectives that will overcome them.** In this step, you are going to verbalize an initial list of obstacles and intermediate objectives. The obstacles are the reasons you can't just snap your fingers in order to see the objectives as a part of your current reality, and the intermediate objectives define what you are going to implement in order to nullify each of the obstacles. While I suggest that you begin with a list of obstacles, it almost doesn't matter what you start with. You will end this step with a scrutinized

* Covey, Stephen R., *The 7 Habits of Highly Effective People,* Simon and Schuster, 1989.

> A mariner must have his eye upon rocks and sands, as well as upon the North Star.
>
> Thomas Fuller, M.D., 1732

list, where every obstacle will have an intermediate objective defined to overcome it, and every intermediate objective will be associated with an obstacle that it is meant to overcome.

a. **List the obstacles to achieving each of the objectives.** An obstacle is an entity that *exists in the current reality* of the system and because of its existence, the objective cannot (or at least there is a perception that it cannot) be achieved. In this step, you are going to succinctly answer the question, *What is preventing us from having the objective/s right now?*.

   It is normal to try to list as obstacles some entities that don't exist now but might in the future. For instance, "The board will say no," or "We won't receive building permits in time." Recognize that these nonexistent entities reflect the predicted effects of presumed, as yet not-taken, courses of action, and that at this stage of the process, you have not planned any course of action! By allowing yourself to include entities that don't yet exist, you are placing obstacles in your own path!

   When such "potential obstacles of the future" come to mind, dig a little deeper to uncover what it is in the current reality that led you to consider them obstacles. For instance, let's consider the two statements, "The board will say no" and "We won't receive building permits in time." A group of condominium owners in Minnesota wants to build a clubhouse and is using the prerequisite tree to define the project plan. It is now late spring, and they are hoping to have the clubhouse built before cold weather sets in. Before anything can happen, the group must present its proposal to the board of their homeowners association. Based on his experience with past proposals to this board, one of the members is predicting, "The board will say no." While this may be a reasonable prediction, as stated it is not an obstacle to getting the clubhouse built. When pressed further to answer *why* he predicts that the board will say no, he came up with the real obstacle: *The board is extremely conservative when it comes to spending the homeowners association's money on anything other than repairs and maintenance.* Another member of the group was concerned about their ability to get building permits in time to get the construction finished by winter. She initially expressed her concern as, "We won't receive the building permits in time." When asked why she predicted this, her response, the real obstacle, was *"Once permits are requested, it takes 4 to 6 weeks to receive them from the county."*

It is also normal to try to list as obstacles some entities that are stated as "the lack of something needed." An example of this type of obstacle would be, "We don't have enough money" or "The building is too small." While this type of obstacle is sometimes unavoidable, try to state it in terms of what *does exist* rather than what doesn't. You might say instead, "We only have $15,000 in the bank," or "The building is 25,000 square feet." You will find in the next steps that defining an intermediate objective for an entity that is clearly stated in terms of what *does* exist is easier than defining an intermediate objective to overcome something that doesn't exist. You will also find that when it comes time to define intermediate objectives, more options will seem to be available when you are dealing with what does exist.

- Here, we bring in some of the tools of sufficient cause thinking to help us define and clarify obstacles — the categories of legitimate reservation. (See Chapter 4.) Check each obstacle for:
  - *Entity existence.* Does the obstacle exist in the current environment? This will help you to avoid the two pitfalls discussed above. Also, if it does not exist, you have just made your implementation plan shorter — no need to work at overcoming a nonexistent obstacle!
  - *Causality existence.* Once you establish that the obstacle is an entity that does exist, use causality existence to check that it really is something that prevents the objective from existing today. You want to be sure that you are dealing with issues that are really preventing you from achieving the objective. If it's not an obstacle, why create a plan to overcome it? If [obstacle], then I am not able to achieve [objective]. Using the causality existence reservation will help you to double check your perceptions, and filter out any entities that are not really obstacles to achieving your objective.

  If you are using the prerequisite tree to plan several objectives, create a list of obstacles for each objective.

Obstacles cannot crush me / Every obstacle yields to stern resolve / He who is fixed to a star does not change his mind.

Leonardo da Vinci, circa 1500

---

b. **For each obstacle, determine an intermediate objective.** In this step you will define an initial list of intermediate objectives — what to implement in order to be able to reach the objectives. Assuming that you are following step 2a with this step, you now have a list (or lists) of obstacles — the main reasons why it is difficult, if not impossible, to achieve the objective/s. Now it's time to determine what you are going to cause to happen, so that the entity will no longer block the objective from being realized. There are two ways to overcome an obstacle:

- Eliminate the entity from reality altogether
- Eliminate the entity's relevance to the objective (achieve the objective *in spite of* the existence of the obstacle)

Take your list of obstacles. Next to each one, write an entity that will, once implemented, overcome the obstacle. An entity that overcomes an obstacle is called an intermediate objective. It must be achieved before the system will be able to achieve its overall objective. It is perfectly OK for one intermediate objective to overcome more than one obstacle. Intermediate objectives must meet the following criteria:

- The obstacle is no longer an obstacle. (The causality reservation again applies here. *If* [intermediate objective], *then* [obstacle] no longer prevents system from achieving [objective].
- You are motivated to put some of your own energy into getting the intermediate objective implemented.
- It's feasible. Intermediate objectives should clearly be easier to implement than the objective itself. If you just can't seem to identify something that seems more feasible to you than the objective, use some of the necessary condition brainstorming techniques described in Chapter 5.

Your intuition may be telling you there is more you must implement if you are going to achieve your objectives. This means that you have additional intermediate objectives — necessary conditions in mind. Scrutinize each using necessary condition thinking, so that you can surface the obstacle/s the intermediate objective is meant to overcome. Each obstacle should meet the criteria outlined in step 2a, and each intermediate objective should meet the criteria outlined in step 2b. Don't be surprised if you throw away one or more of these necessary conditions as you scrutinize them. Often times, we create much longer implementation plans than necessary, because we assume that we must do things that we don't need to do.

Let's apply the first two steps to a small example. The president of a small company has an objective to increase his engineering staff by 12 engineers. She has decided to do a prerequisite tree to map out her basic plan. She has stated the objective of the prerequisite tree to be *We have 12 new engineers.*

The left side of the Table 10.1 represents the obstacles defined in step two. The right side of the table represents the intermediate objectives that she selected to overcome the corresponding obstacle.

Table 10.1

Obstacle ***(Entity that exists in current reality which blocks the objective.)***	***Intermediate Objective*** ***(Entity that, once implemented, overcomes the obstacle.)***
There is a hiring freeze in our company.	The hiring freeze is lifted.
There is a scarcity of qualified engineers in our area.	We conduct a national search.
The ongoing rate for engineers in the current market is much higher than what our existing engineers earn.	Our existing engineers get increases to match the going market rates.

3. **Map the implementation order of the intermediate objectives.** In step 2a you identified the obstacles to achieving the objective/s of the tree (plan). These are the reasons that something must be done, if you hope to achieve your objectives by a means other than luck. In step 2b, you made *decisions*. You selected intermediate objectives — conditions to be achieved, or actions to be taken, that once accomplished, will remove the obstacles, enabling you to move that much closer to achieving the objective/s. You have created new necessary conditions — you must accomplish each of the intermediate objectives in order to accomplish the prerequisite tree's objectives/s, because of the existence of the

obstacles. Now it is time to determine the order in which the intermediate objectives must be accomplished. What should be worked on first, second, and third? Is there an inherent order among the intermediate objectives? Must any of the intermediate objectives come before the others? Can any of the intermediate objectives be accomplished in parallel? You will answer these questions by establishing more precisely any necessary condition relationships that exist among the intermediate objectives.

a. **Identify two intermediate objectives that appear to have a time dependency between them.** Your intuition is telling you that one needs to be implemented before or after the other (Figure 10.3).
b. **Diagram the relationship as a necessary condition relationship.** Between the two, the arrow points from the earlier intermediate objective to the later one. The obstacle that the intermediate objective overcomes is verbalized on the arrow itself.
c. **Scrutinize the relationship you've just diagrammed using necessary condition thinking** (see Chapter 5), adding intermediate objectives and obstacles if necessary. In this step, you are checking the assumptions behind the need to accomplish intermediate objective io-1 before io-2.
   i. The first element you are going to check is the obstacle ob-1 (see Figure 10.4). Is this entity preventing the accomplishment of intermediate objective io-2? If so, when you read the necessary condition relationship, it will make sense and ring true. Check that the obstacle is an obstacle to the intermediate objective it now points to, using the causality existence reservation:
      - We cannot accomplish [intermediate objective io-2], because of the existence of [obstacle ob-1].
      - If [obstacle ob-1], then we are unable to accomplish [intermediate objective io-2].

      If you are satisfied that obstacle ob-1 does, in fact, block intermediate objective ob-2, move on to step 4cii.

      If you are not satisfied that obstacle ob-1 blocks intermediate objective io-2, and still believe that intermediate objective io-1 must be in place before io-2, then ask yourself why. It could very well be that there is another obstacle preventing io-2 from existing, and your intuition is telling you that io-1 will overcome that obstacle as well as obstacle ob-1. Verbalize the new obstacle by asking any of the following questions:

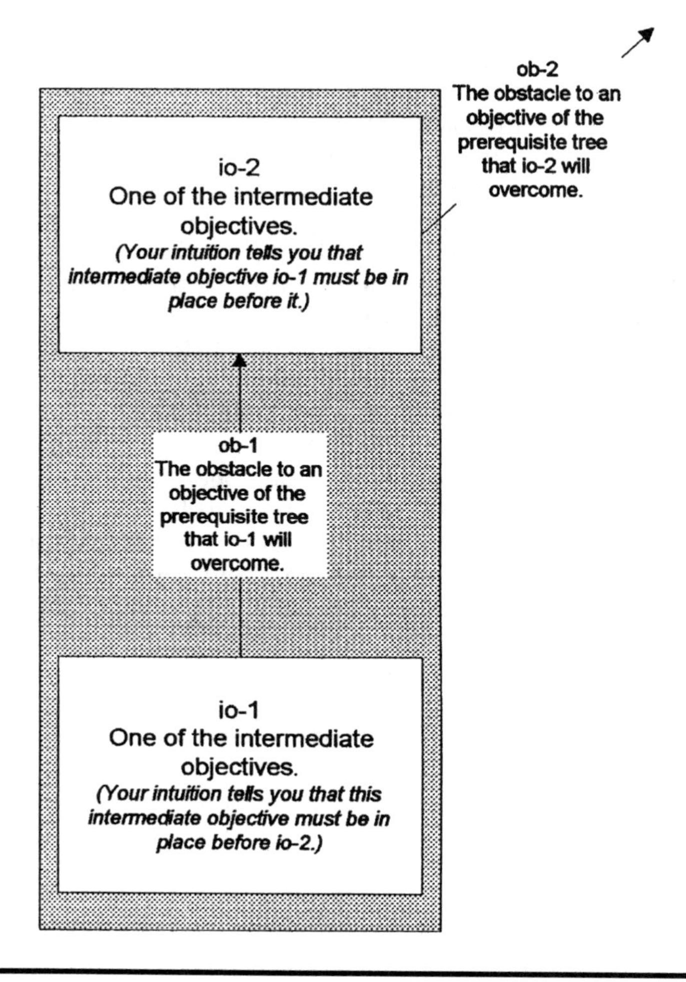

Figure 10.3

- *Why do I believe that io-1 must come before io-2?*
- *We must implement io-1 before io-2 because [new obstacle].*
- *We cannot achieve io-2 without first implementing io-1 because [new obstacle].*

If you do verbalize a new obstacle, test it as an obstacle to io-2, using the criteria defined in step two. Your diagram may now look something like Figure 10.4.

If you are unable to verbalize a new obstacle, it is quite possible that you have just discovered that there is not a real dependency between the two intermediate objectives.

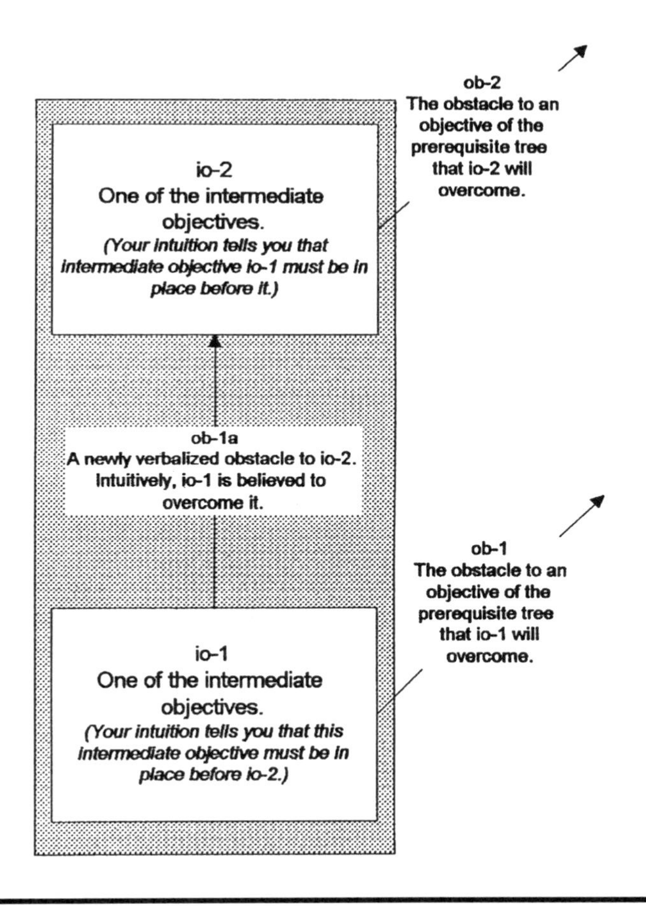

Figure 10.4

That's OK! It's simply a reflection of our habit of trying to think and plan in a linear fashion. First we do this, then we do that, etc. The prerequisite tree process has us rigorously examine and document only the necessary order. When we understand the whys behind the musts, it is easier to make the decisions that will help us focus on "first things first." Go back to your list, select another pair, and start again at step 3a.

ii. Now you are going to check the necessary condition relationship between the two intermediate objectives. Does intermediate objective io-1 clear the obstacle from the path to intermediate objective io-2? Use necessary condition thinking to validate the relationship:

- In order to accomplish [intermediate objective io-2], we must accomplish [intermediate objective io-1], because [obstacle ob-1].
- When we accomplish [intermediate objective io-1], then [obstacle ob-1] no longer prevents us from accomplishing [intermediate objective ob-2].

If it makes sense, and rings true for your situation, move on to step 3d. If not, it is likely that you need to add an intermediate objective that must be implemented between io-1 and io-2. What, once implemented, will overcome the obstacle to io-2? Define the new intermediate objective, using the guidelines described in step 2b. Insert the new intermediate objective in the diagram and scrutinize according to the guidelines in this step. It is important to note that it is possible that an intermediate objective from your existing list is appropriate. Finally, scrutinize the new connection you have created between the original intermediate objective io-1 and the new io-3 (Figure 10.5).

d. Select another intermediate objective from your list that appears to have a time dependency with any of the intermediate objectives already diagrammed. Looking at our generic sample, you select one that:
   - Must be implemented before io-1, or
   - Must be implemented after io-1, or
   - Must be implemented before io-2, or
   - Must be implemented after io-2, or
   - Must be implemented before io-3, or
   - Must be implemented after io-3.

   Add the selected intermediate objective to the tree as a necessary condition diagram that illustrates the dependency between it and the intermediate objective with which it shares the time dependency. Scrutinize this relationship according to the guidelines provided in step 3a through 3c.

Let's return again to the president of the company who was using the prerequisite tree to determine the path to having 12 new engineers on board. It was immediately obvious that before she could implement a national search, she would need to lift the hiring freeze she had imposed the previous year. So, the first two intermediate objectives she selected were "The hiring freeze is lifted" and "We do a national search." Figure 10.6 shows the results of steps 3a through 3c. Figure 10.7 illustrates her prerequisite tree after step 3d.

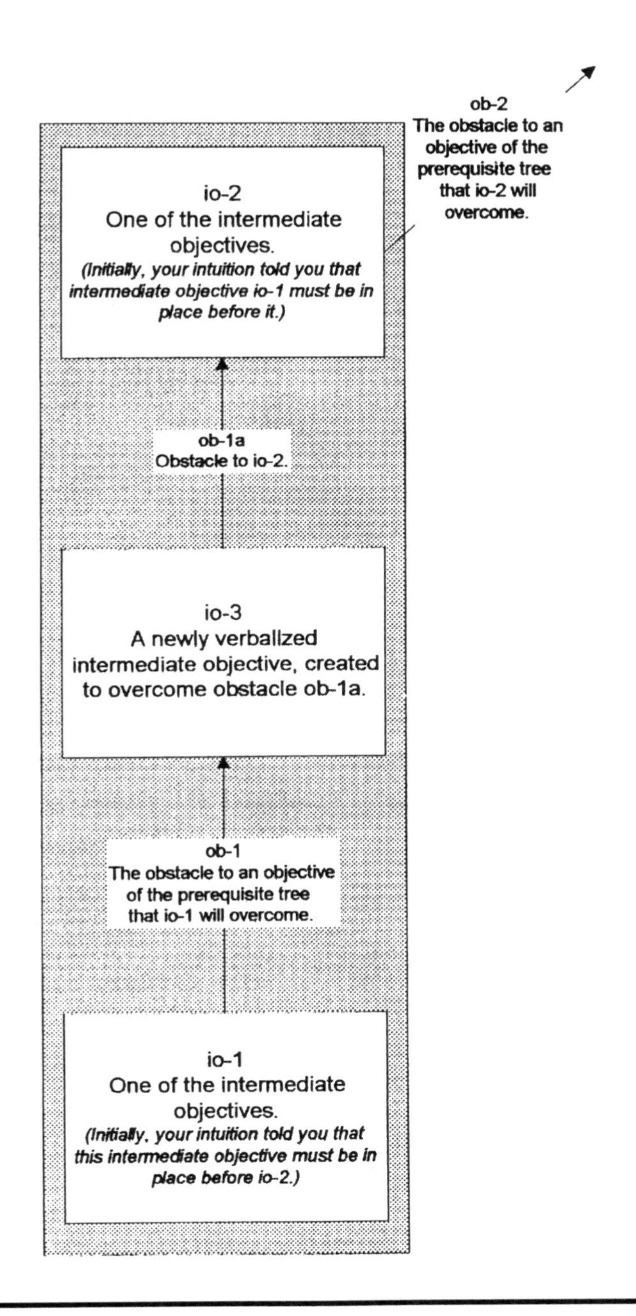
ob-2
The obstacle to an objective of the prerequisite tree that io-2 will overcome.
io-2
One of the intermediate objectives.
(Initially, your intuition told you that intermediate objective io-1 must be in place before it.)
ob-1a
Obstacle to io-2.
io-3
A newly verbalized intermediate objective, created to overcome obstacle ob-1a.
ob-1
The obstacle to an objective of the prerequisite tree that io-1 will overcome.
io-1
One of the intermediate objectives.
(Initially, your intuition told you that this intermediate objective must be in place before io-2.)

Figure 10.5

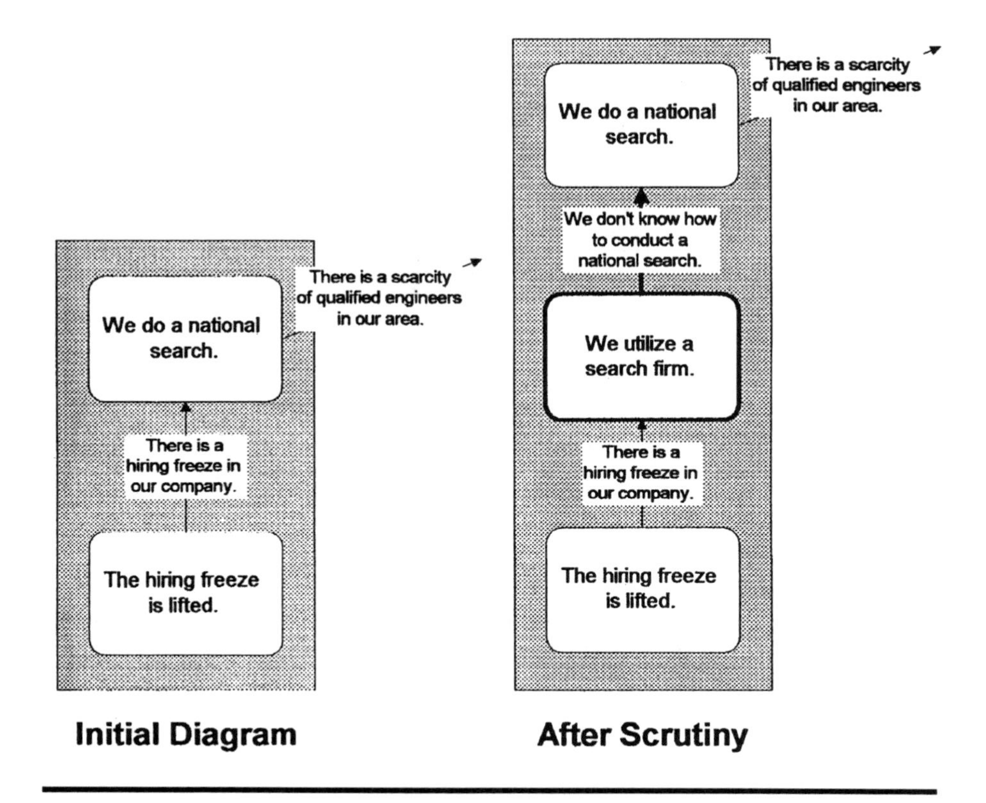

Figure 10.6

e. **Repeat step 3d until no more of the intermediate objectives from your list can be connected.**

f. **Connect the cluster, any remaining entities, and the objective.** Now you have a cluster of necessary condition relationships. At the top of the cluster are one or more intermediate objectives, each of which is still required in order to overcome obstacles to the overall objective of the prerequisite tree. Attach an arrow from each of these "top" intermediate objectives to the overall objective. Scrutinize the relationships as you did in step three.

- You may also have some intermediate objectives that are not part of the larger cluster. These are the intermediate objectives that have no necessary dependencies with the intermediate objectives in the cluster. For these,
  - Establish any dependencies among them, according to the step three guidelines.

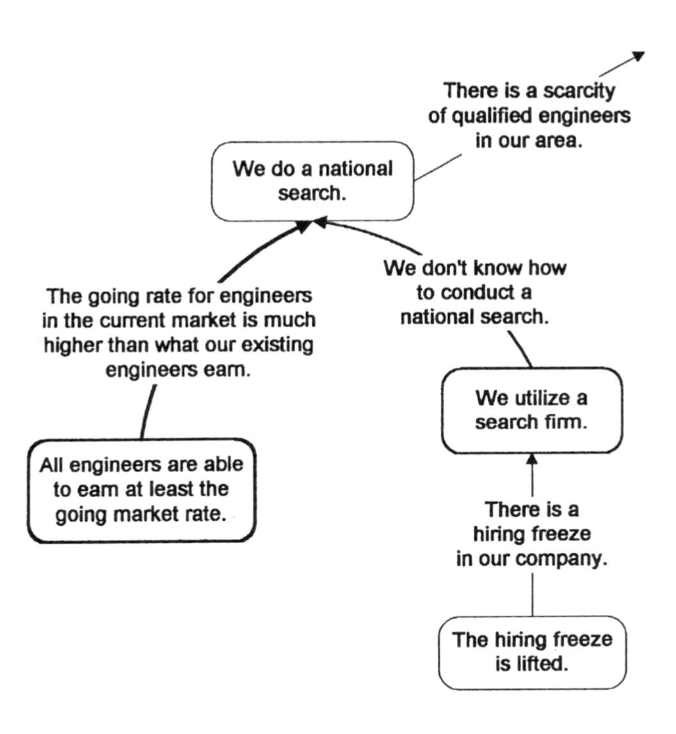

Figure 10.7

- Connect the top of any newly formed clusters to the overall prerequisite tree objective as described above.
- Connect the top of any single intermediate objectives that have no necessary dependencies with any cluster to the overall prerequisite tree objective as described above.

- **When the prerequisite tree has multiple objectives,** you will likely find the following:
  - Several of the objectives will be intermediate objectives for other objectives.
  - Many of the intermediate objectives established initially for one objective are also intermediate objectives for other objectives.

  Start by creating the prerequisite tree for one of the objectives, preferably the one that appears to be the most difficult to achieve, or the one that you know intuitively will take the longest to accomplish. Most of the other objectives, including their obstacles and intermediate objectives, will likely be a part of that prerequisite tree. It will then be easier for you to add any remaining objectives, intermediate objectives, and obstacles.

g. **If necessary, build down to ensure that every entry point is actionable.** Every entry point (entity that has no arrows pointing into it) to the prerequisite tree should be an entity that can be accomplished somewhat easily. When you look at an entry point, you should be able to say, "I know how we can do that!" Action should be coming to mind. You should be able to "just do it" or know enough to create an action plan (such as a transition tree). For any entry point that is not easily accomplished, repeat the basic process:
   - Determine any obstacles that prevent its attainment (see guidelines in step 2a)
   - Select an intermediate objective for each obstacle (see guidelines in step 2b)
   - Map them, using necessary condition thinking, to the entry point (see guidelines in step 3)

The final version of the small company president's "Twelve New Engineers" prerequisite tree looks like Figure 10.8.

It is important to note that obstacles depend to a large degree on who's doing the tree. For instance, if the engineering manager of this firm had been creating the prerequisite tree, he may have identified several obstacles to the entry point, "We establish a throughput-based compensation system." The president of the firm, however, didn't see it that way. She knew it was within her power to make that intermediate objective happen, and didn't see any major hurdles to getting such a system in place.

A few years ago, I helped provide training in the thinking processes to several groups of people in the Department of Defense. Every once in a while, someone would have an idea, and the response to the idea would be the statement of a huge obstacle, "Yea, right... that would take an act of Congress!" The typical next step would be to search for a more feasible idea. One of the groups, however, was a group of generals, led by Dale Houle of the Goldratt Institute. Some ideas that were generated also had the same obstacle: "Yea, right... that will take an act of Congress!" However, with that group, the comment was often followed by, "OK, so which one of us should be assigned that item?"

4. **Implement!** With the prerequisite tree, you have made decisions on the path/s that must be taken in order to get from where your system is today, to where you want it to be in the future. You have determined what needs to happen, why it needs to happen,

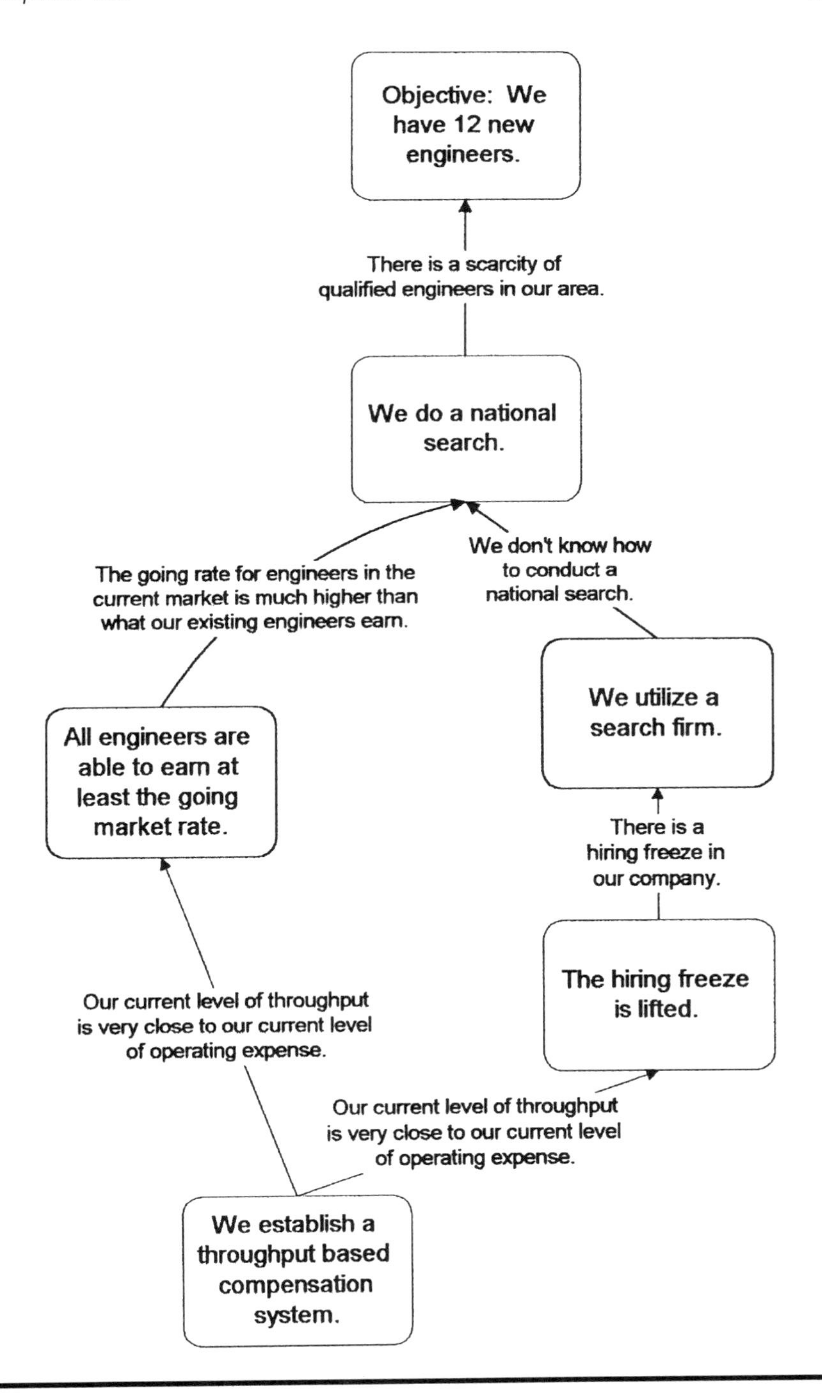
Objective: We have 12 new engineers.
There is a scarcity of qualified engineers in our area.
We do a national search.
The going rate for engineers in the current market is much higher than what our existing engineers earn.
We don't know how to conduct a national search.
We utilize a search firm.
All engineers are able to earn at least the going market rate.
There is a hiring freeze in our company.
The hiring freeze is lifted.
Our current level of throughput is very close to our current level of operating expense.
Our current level of throughput is very close to our current level of operating expense.
We establish a throughput based compensation system.

Figure 10.8

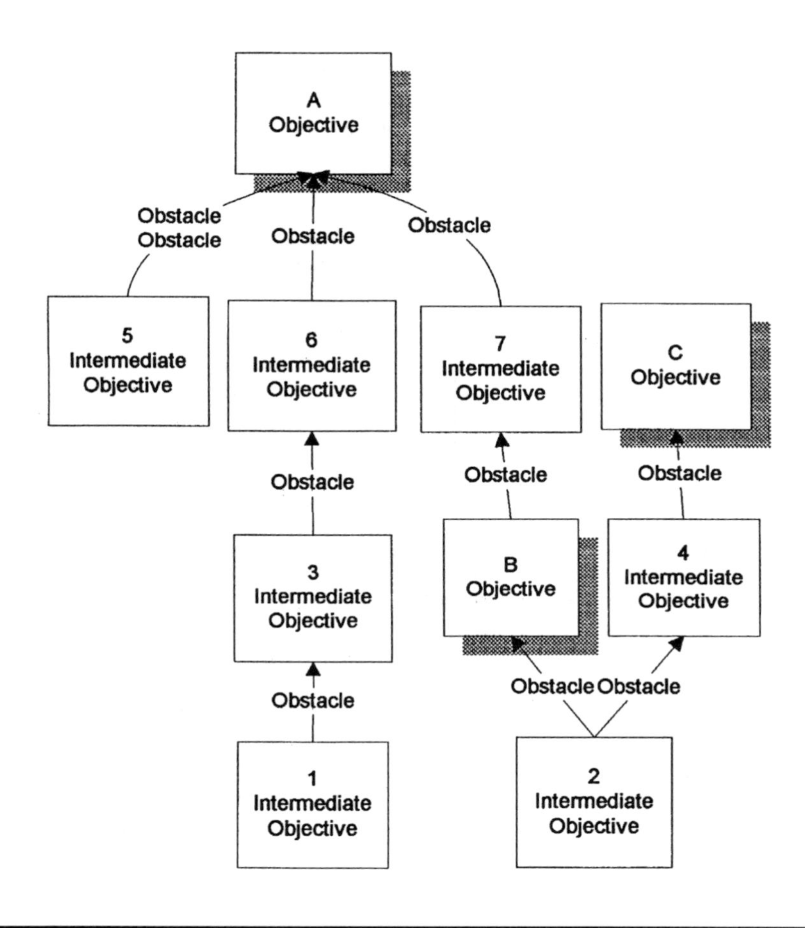

Figure 10.9

and in what order things must happen if you are to achieve your objective/s. It is now time to take the first steps and make it happen. Many prerequisite trees have more than one entry point. How do you know what to start first? Let's take a look at the generic prerequisite tree for some guidance (Figure 10.9).

In the prerequisite tree, we have defined the major milestones that must be accomplished, and we have established the necessary dependencies that link the milestones to each other and the objectives. This should be the first step of any major project plan. However, before we can answer the question, "What should we work on when," we need some additional information:

- What resources are required in order to accomplish each of these milestones and when are those resources available?
- How much time will it take each of those resources to do the tasks that they need to do in order to accomplish the milestones?

Most planning and implementation methodologies do not take all of these factors into account (necessary dependencies, necessary resources, necessary time). The absence of any will lead to implementations that take longer and cost more than they are predicted to take. *Critical chain* is the only method for project planning that I have run across which considers all. Two excellent books for learning more about this superb approach are *Critical Chain** by Dr. Eliyahu Goldratt, and *Project Management in the Fast Lane: Applying the Theory of Constraints*** by Rob Newbold.

Finally, as you approach the time to implement each of the intermediate objectives, you may choose to create an action plan. The transition tree (see Chapter 6) is an excellent tool for doing so, where one or more intermediate objectives serve as the objectives for the transition tree.

Example

The following two pages (Figures 10.10A through D) represent a portion of a prerequisite tree that was prepared by a team of managers that were responsible for growing the market share and profitability of a business unit of a large multinational corporation. A current reality tree, evaporating cloud, and future reality tree preceded the prerequisite tree. They used the prerequisite tree to determine what needed to happen in order to implement the injections established in the future reality tree. The boxes that are shadowed are some of the injections.

* Goldratt, Eliyahu M., *Critical Chain,* North River Press, 1997.

** Newbold, Rob, *Project Management in the Fast Lane: Applying the Theory of Constraints,* St. Lucie Press, 1998.

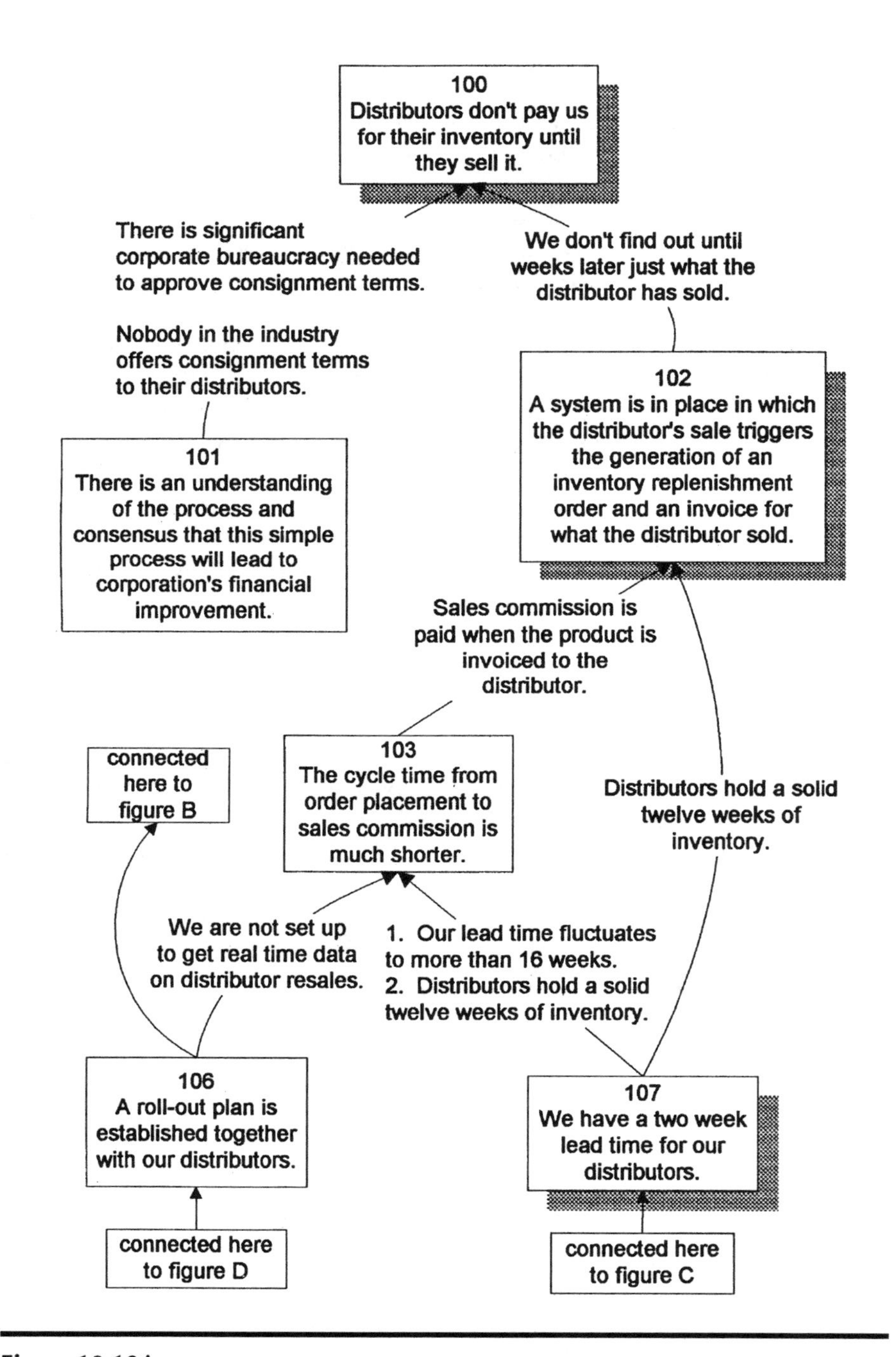

Figure 10.10A

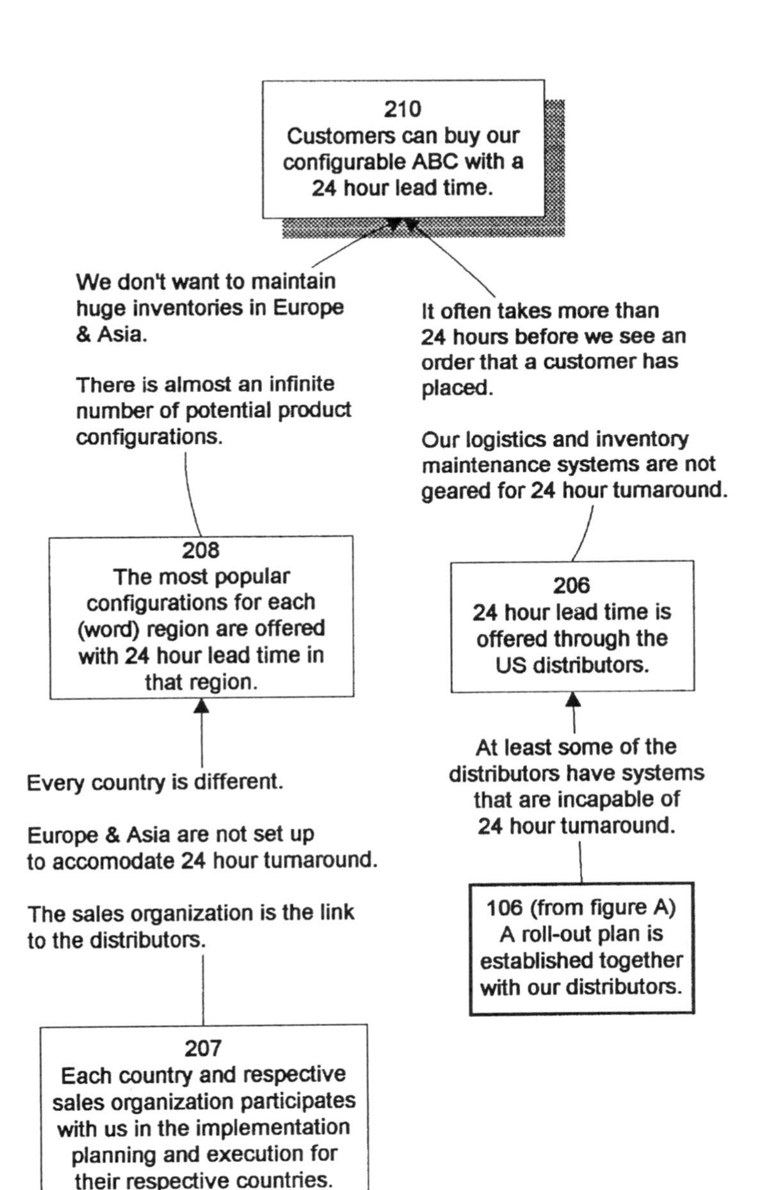

Figure 10.10B

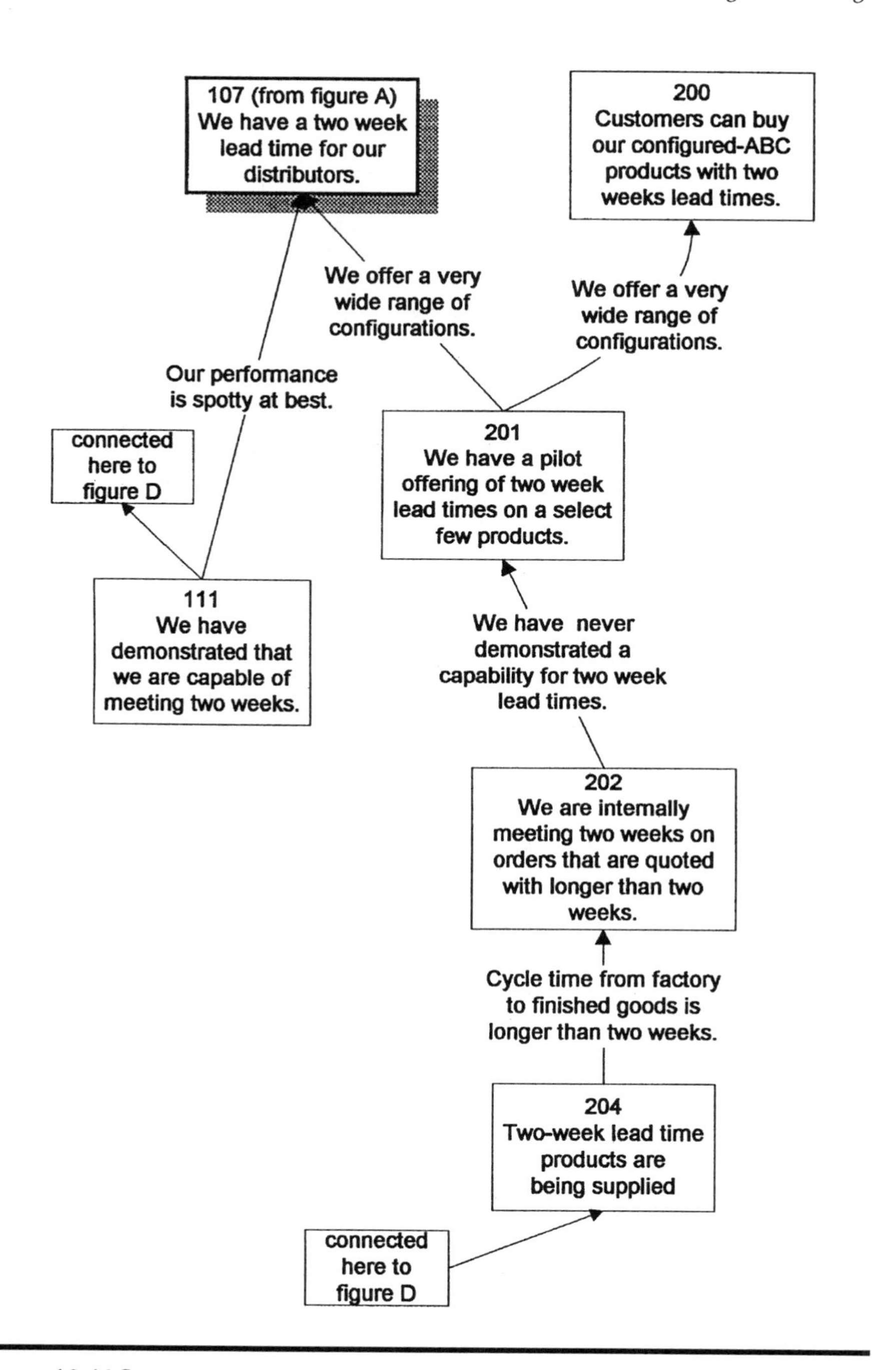

Figure 10.10C

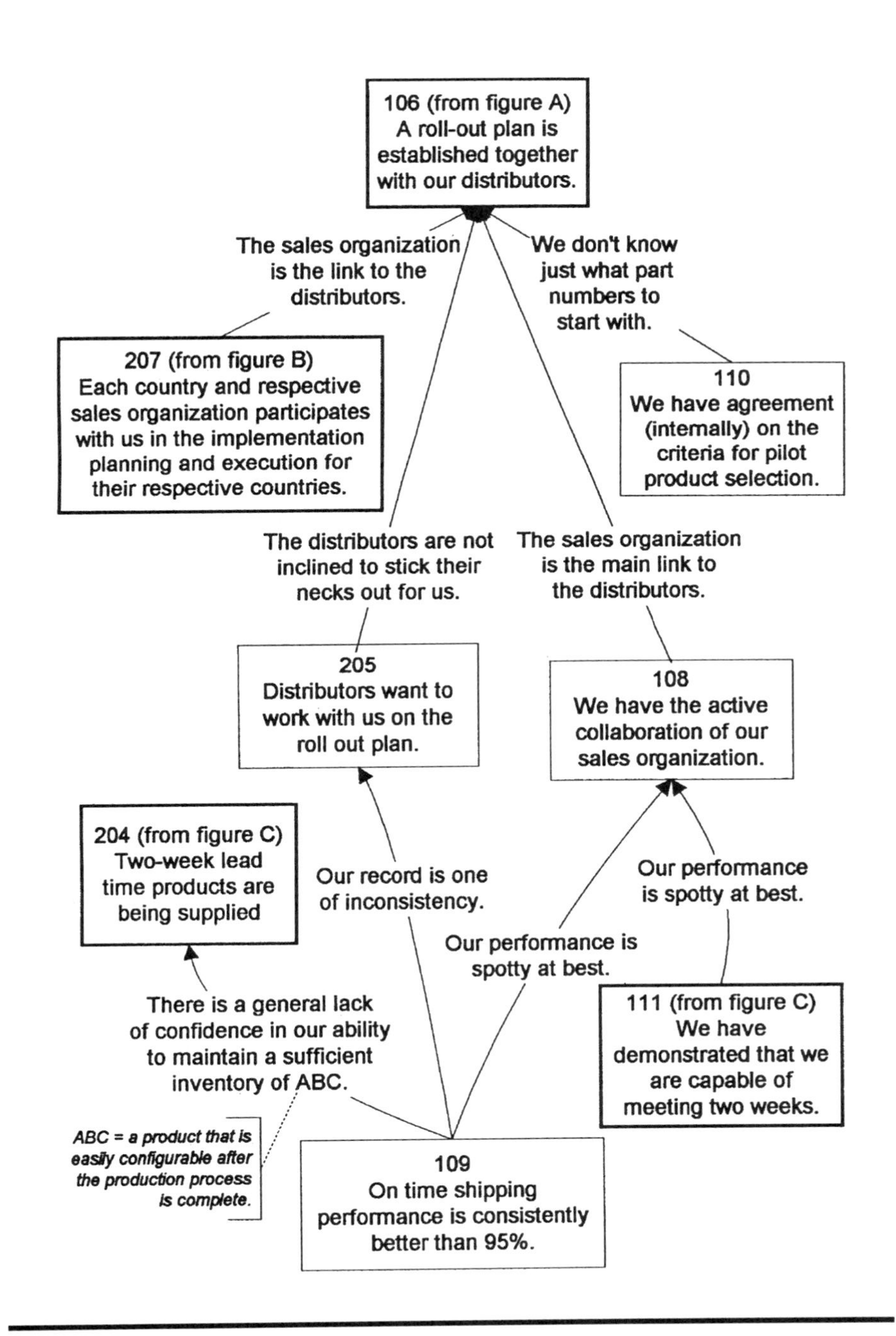

Figure 10.10D

Part Three

Chapter 11

The Full Analysis

> If you work on the fly, you do things fast. But you may do the wrong things —which slows down the project . ... My approach is 50% planning, 25% doing, and 25% testing and training. It's a magic formula around here.*
>
> Chris Higgins
> Bank America, 1998

The Full Thinking Process Analysis uses all five application tools to analyze a system or situation in order to identify a core problem, develop a solution for it, and determine how to implement that solution. In other words, the full analysis answers the three basic questions that any system should answer when going about the task of improving itself:

- *What* to change?
- *To what* to change?
- *How to cause* the change?

Figure 11.1 illustrates the three questions surrounding change and the role that each of the thinking processes will play in answering the questions when you utilize the full analysis.

* *Fast Company Magazine,* Boston, June/July 1998, issue 15.

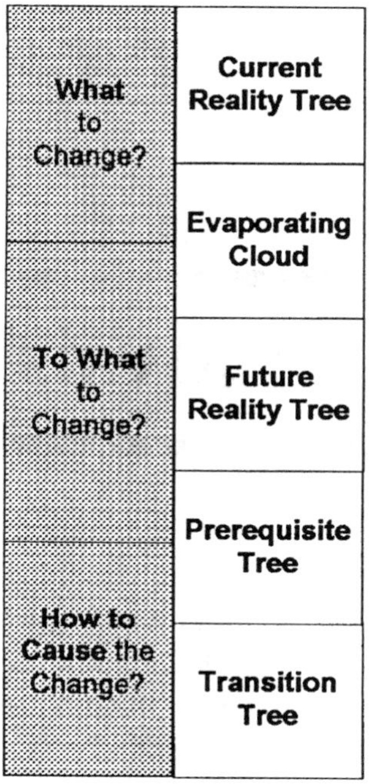

What to Change?	Current Reality Tree
	Evaporating Cloud
To What to Change?	Future Reality Tree
	Prerequisite Tree
How to Cause the Change?	Transition Tree

Figure 11.1

Once you have finished doing such a full analysis, you won't need convincing that each step of the way has built-in safety nets. What you miss in one step, you will likely catch in the next. While the task of systematically utilizing each of the five application tools in sequence may seem like quite a bit of analysis and planning, chances are that you will, as a result, develop and implement more robust solutions that *really* address the constraints of the system.

Some ways in which the full analysis is utilized are to create strategic plans, operations plans, market strategies, and general improvement plans.

Overview of the Process

1. Create a current reality tree to identify a core problem.
2. Create an evaporating cloud to identify a systemic conflict that is perpetuating the core problem, to brainstorm solutions to the core problem, and to select the initial elements of the solution.
3. Using an injection selected in step 2 as your starting point, create a future reality tree to develop a robust solution to the core

problem, eliminating the undesirable effects and blocking undesirable consequences.

4. Create a prerequisite tree to determine the necessary conditions (intermediate objectives) for implementing the injections (objectives), and the sequence in which they should be accomplished.
5. Create transition trees to define the specific action plans for accomplishing the intermediate objectives and objectives of the prerequisite tree.

This is the order in which the thinking processes were taught when they were first introduced. A downside to the teaching approach was that many students of the thinking processes developed a paradigm that said, "Thou must always use these thinking processes, beginning with the current reality tree and continuing on through completion of the transition trees." This led to the belief that using the thinking processes is always a long, laborious task. As you have seen throughout this book, that is not the case.

It is worth repeating one of my main messages here — use the tools to answer questions that you have yet to answer. If you are convinced that you know the problem, if you are convinced that you know the solution, if you are convinced that you know exactly what to do, don't go for the full analysis. Use the tools you need to answer the questions you need to answer. As with picking up any of the single application tools, when you embark on a full TP analysis, make sure you are willing to learn. Make sure you are willing to recognize, verbalize, challenge, and change your assumptions.

The Process in Detail

1. **Current Reality Tree.**
   By articulating the undesirable effects, the causalities that exist among them, and a core problem that serves to keep the undesirable effects in existence, the current reality tree will answer the question, "What to change?" Follow the process outlined in Chapter 8 to create a current reality tree and identify a core problem.
2. **Evaporating Cloud.**
   In Chapter 9, we established that every problem is, in fact, a conflict. Thus, we should be able to articulate every core problem as a conflict. Let's call this conflict a core conflict. The evaporating cloud will be used to verbalize the core problem as a systemic conflict that is perpetuating the existence of the undesirable effects; brainstorm solutions to this core conflict; and select the initial elements of the solution. As illustrated in Figure 11.2, use the following to create the evaporating cloud diagram:

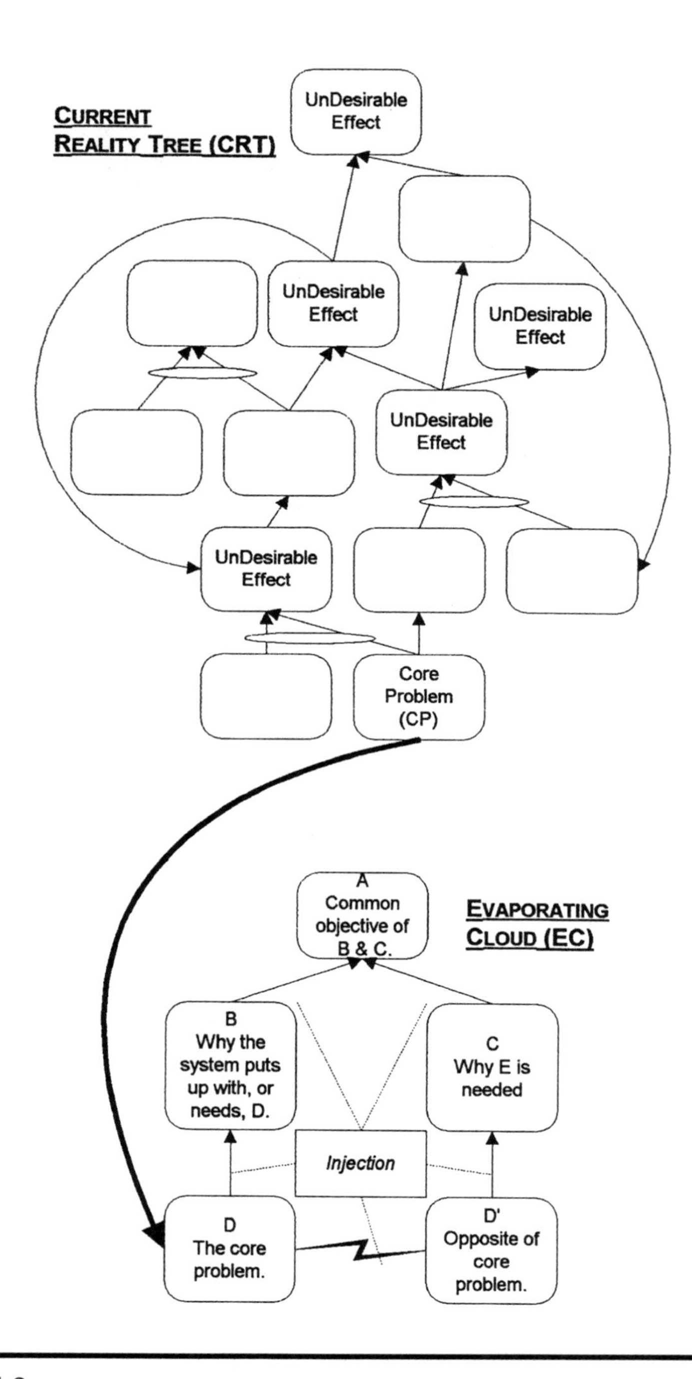
CURRENT
REALITY TREE (CRT)
UnDesirable Effect
UnDesirable Effect
UnDesirable Effect
UnDesirable Effect
UnDesirable Effect
Core Problem (CP)
A
Common objective of B & C.
EVAPORATING
CLOUD (EC)
B
Why the system puts up with, or needs, D.
C
Why E is needed
Injection
D
The core problem.
D'
Opposite of core problem.

Figure 11.2

a. The D and D′ entities are simply the core problem and its opposite.
   - The D entity is the core problem. Copy that entity from your current reality tree.
   - The D′ entity is the opposite of the core problem. *What entity should replace the core problem in the future?*
   - Read these two entities back to yourself. The D entity is one you don't want to exist, because it is a core problem. The D′ entity is one that you do want to exist, because the opposite of the core problem will result in the opposite of many undesirable effects. D and D′ cannot coexist in the system as it exists today.

b. Now it's time to determine why these entities are needed in the system by defining what they are the necessary conditions of.
   - Face it, the core problem is an entity that does exist in the current reality. If there were no real or perceived need for this entity to exist, it wouldn't. Its existence is satisfying some need of the system, and the B entity of the evaporating cloud articulates that need. What is it? *Why does the system need to continue to put up with [D]?* Or, *What need of the system does putting up with the core problem (D) fulfill?*
   - The C entity answers the question, *What need of the system will [D′ entity] satisfy?* Or, *Why does the system need [D′]?*

c. The A entity answers the question, *Why does the system need both [B] and [C]?* This is often a global objective of the system. You might find it to be the subject of your current reality tree, the purpose of the system, or a key aspect of what forms the system's identity.

The transition from "*What* to Change" to "*To What* to Change" occurs during the evaporating cloud process. When you have constructed the cloud, you have completed your answer to the "*What* to Change" question. *It is this conflict that must change!* You will begin to answer the "*To What* to Change" once you have surfaced assumptions, brainstormed injections, and selected an injection with which to begin to create the solution.

After following the rest of the process for evaporating clouds as defined in Chapter 9, select an injection. The injection should meet the following criteria:

- You really like the idea.
- You are willing to invest the time to ensure its implementation.
- You see that there can be a (cause–effect) path from the injection to the removal of the core problem.

Don't worry too much about obstacles to implementing the injection or potential negative effects of the injection. The future reality and prerequisite trees will take care of those issues. (However, if you have several injections that meet the criteria listed above, by all means, select the more practical of the bunch.)

3. **The Future Reality Tree**
   Now comes the task of creating your vision of the future — the replacement for the current reality. You will create a future reality tree in order to formulate the answer to the question, "*To What* to Change?" The injection that you selected at the conclusion of the evaporating cloud process is the injection you will use to start your future reality tree. Remember that the future reality tree process calls for you to list the positives and negatives of the injection. The positives of the injection (the objectives of the future reality tree) are the opposites of the current reality tree's reselected undesirable effects.
   - For each reselected undesirable effect, determine its opposite. "Opposite" should answer the question, *What entity do you want to see in the future instead of this undesirable one?* Write each as an entity, stated as if it exists in the present tense.
   - Build the future reality tree as described in Chapter 7. The future reality tree is complete when all of the objectives (opposites of the current reality tree's reselected undesirable effects) are achieved, all potential negative branches (that you can think of) have been removed, and at least one strong positive reinforcing loop has been created (Figure 11.3).
4. **Prerequisite Tree**
   You will use the prerequisite tree to create your implementation road map. Create the prerequisite tree using the process described in Chapter 10. The injections from your future reality are the objectives of the prerequisite tree.

   The intermediate objectives formulate the rest of the answer to the question, "*To What* to Change?" The completed prerequisite tree, the ordered sequences for implementation that are created by utilizing necessary condition thinking, provides the first part of the answer to "How to cause the change" (Figure 11.4).
5. **Transition Tree**
   Create transition trees when you are ready to define specific, detailed action plans for parts of or all of the implementation. The intermediate objectives and injections defined on the prerequisite tree are the objectives for your transition trees. This is often the

point at which delegation occurs. Managers will delegate one or more intermediate objectives or injections to their employees or teams, who then go on to create the transition trees that describe how they intend to accomplish those objectives. This provides a great document for the manager to review with the employee or team before, during, and after they proceed with action. The manager can verbalize his or her concerns, focusing on the causalities or potential consequences of what's been designated on the tree. The result is an action plan that is supported by the manager and truly owned by the employee/team. (Figure 11.5).

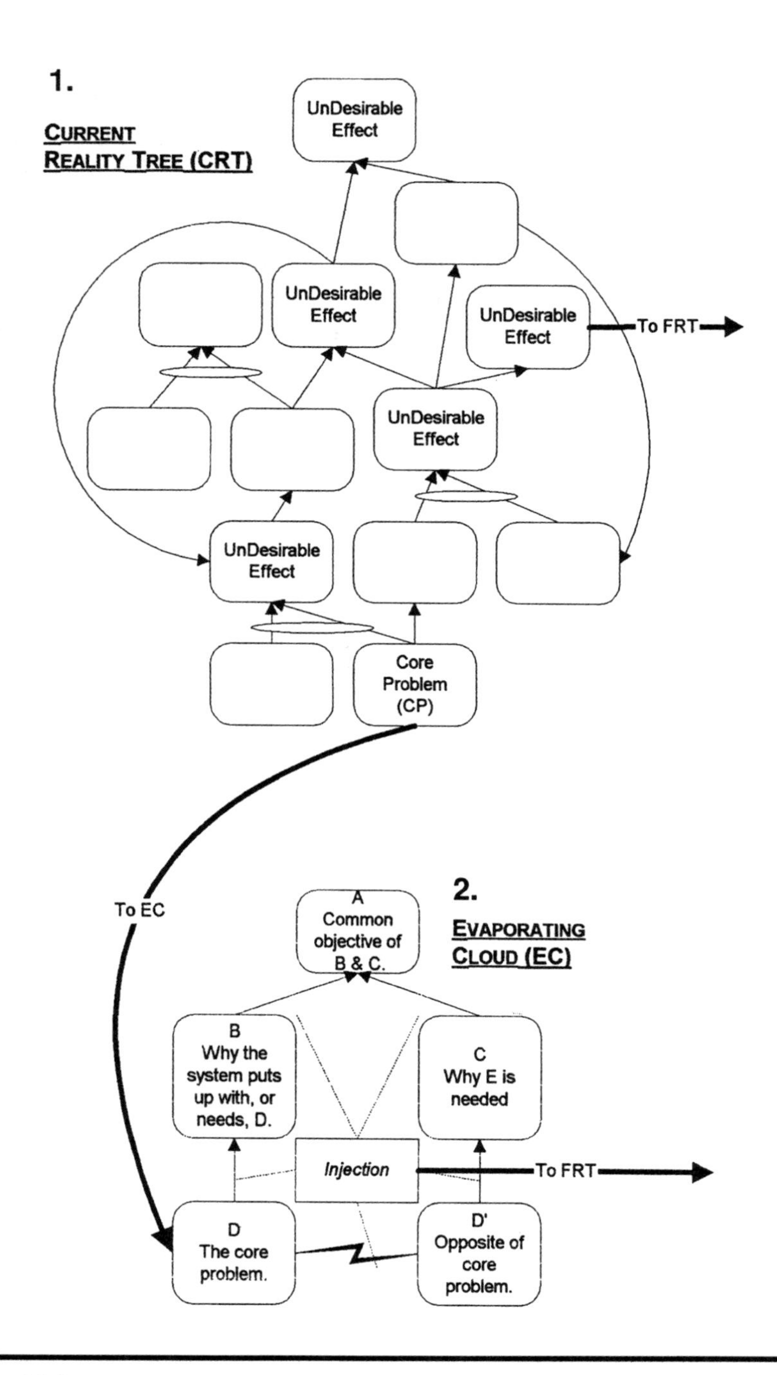

1.
CURRENT REALITY TREE (CRT)
UnDesirable Effect
UnDesirable Effect
UnDesirable Effect
To FRT
UnDesirable Effect
UnDesirable Effect
Core Problem (CP)
To EC
2.
EVAPORATING CLOUD (EC)
A Common objective of B & C.
B Why the system puts up with, or needs, D.
C Why E is needed
Injection
To FRT
D The core problem.
D' Opposite of core problem.

Figure 11.3a

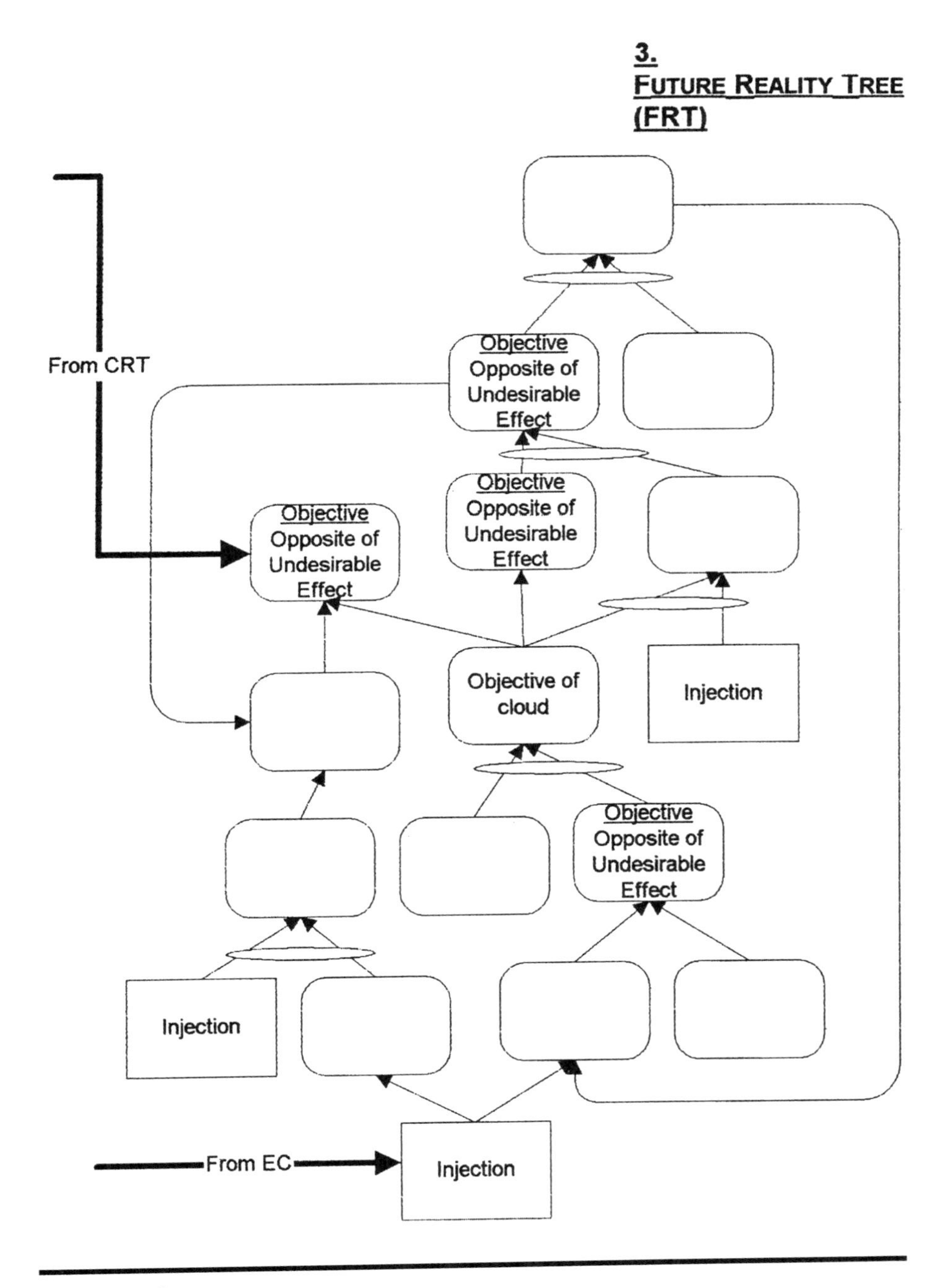
3.
FUTURE REALITY TREE
(FRT)
From CRT
Objective
Opposite of
Undesirable
Effect
Objective
Opposite of
Undesirable
Effect
Objective
Opposite of
Undesirable
Effect
Objective of
cloud
Injection
Objective
Opposite of
Undesirable
Effect
Injection
From EC
Injection

Figure 11.3b

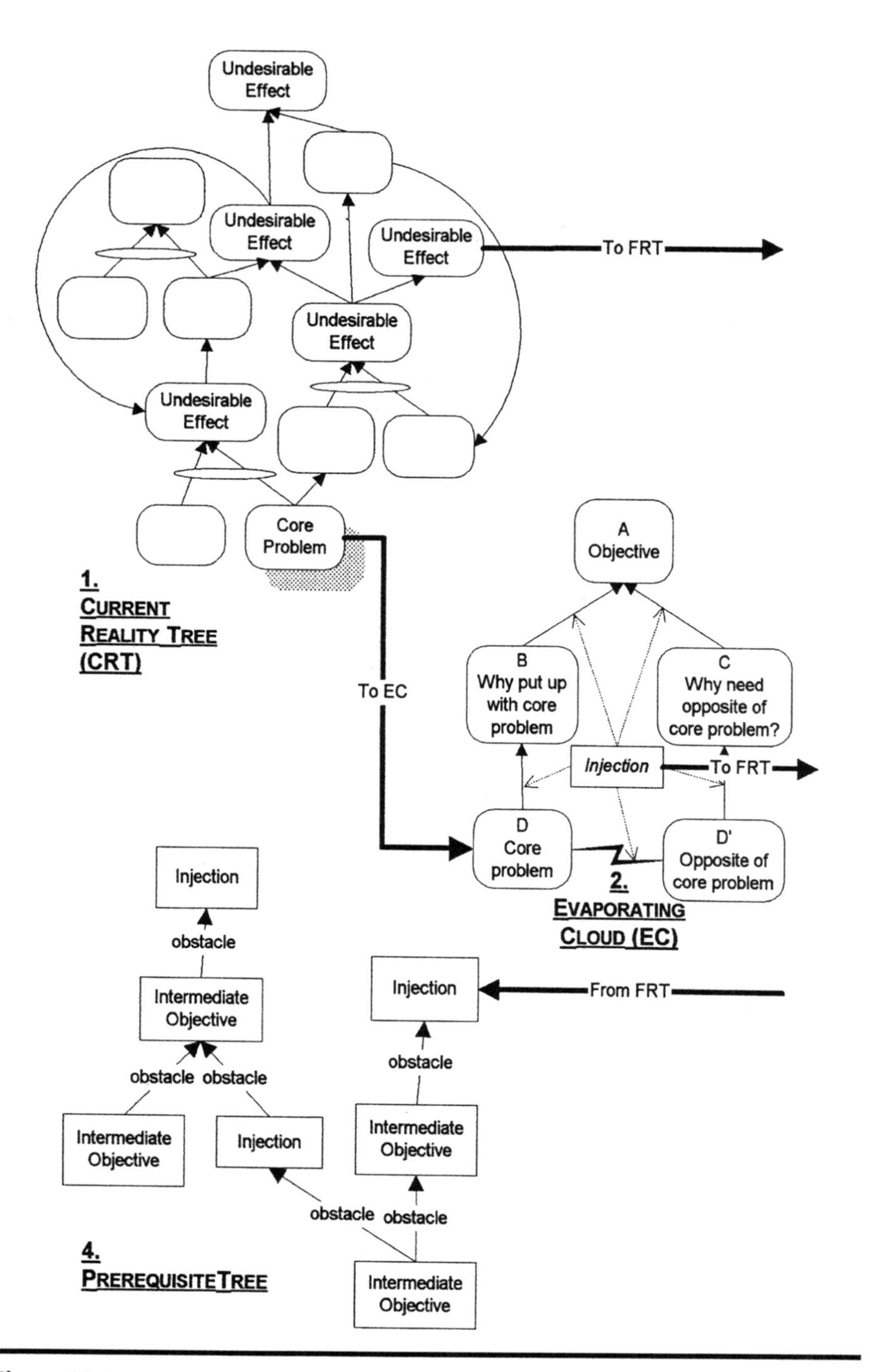
Undesirable Effect
Undesirable Effect
Undesirable Effect
To FRT
Undesirable Effect
Undesirable Effect
Core Problem
1. CURRENT REALITY TREE (CRT)
To EC
A Objective
B Why put up with core problem
C Why need opposite of core problem?
Injection
To FRT
D Core problem
D' Opposite of core problem
2. EVAPORATING CLOUD (EC)
Injection
obstacle
Intermediate Objective
obstacle obstacle
Intermediate Objective
Injection
Injection
From FRT
obstacle
Intermediate Objective
obstacle obstacle
4. PREREQUISITE TREE
Intermediate Objective

Figure 11.4a

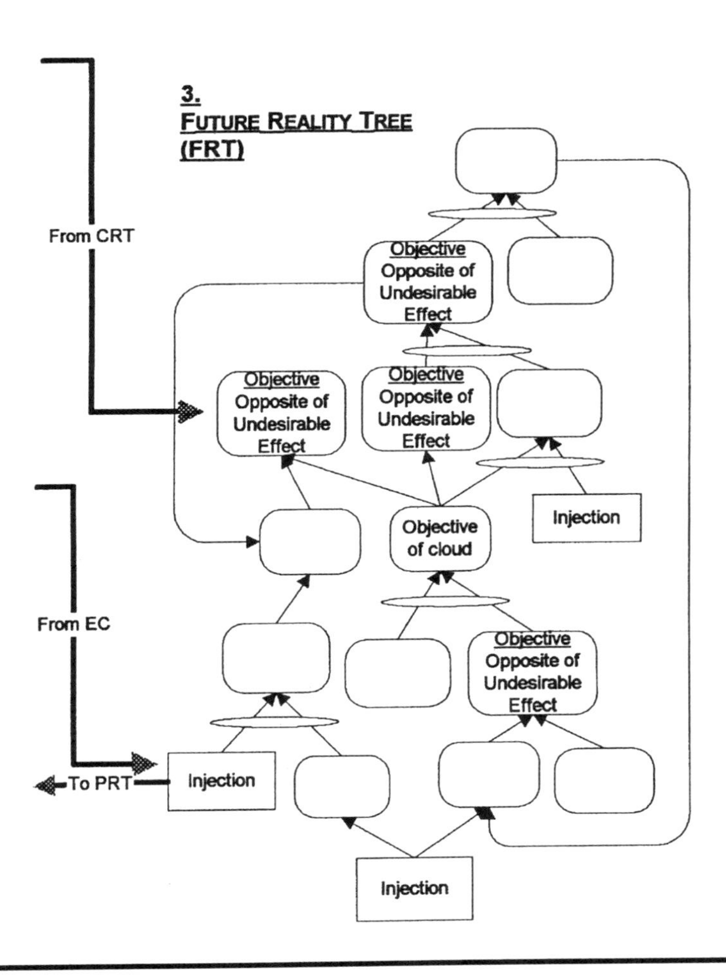
From CRT
3.
FUTURE REALITY TREE
(FRT)
Objective
Opposite of
Undesirable
Effect
Objective
Opposite of
Undesirable
Effect
Objective
Opposite of
Undesirable
Effect
Injection
Objective
of cloud
From EC
Objective
Opposite of
Undesirable
Effect
To PRT
Injection
Injection

Figure 11.4b

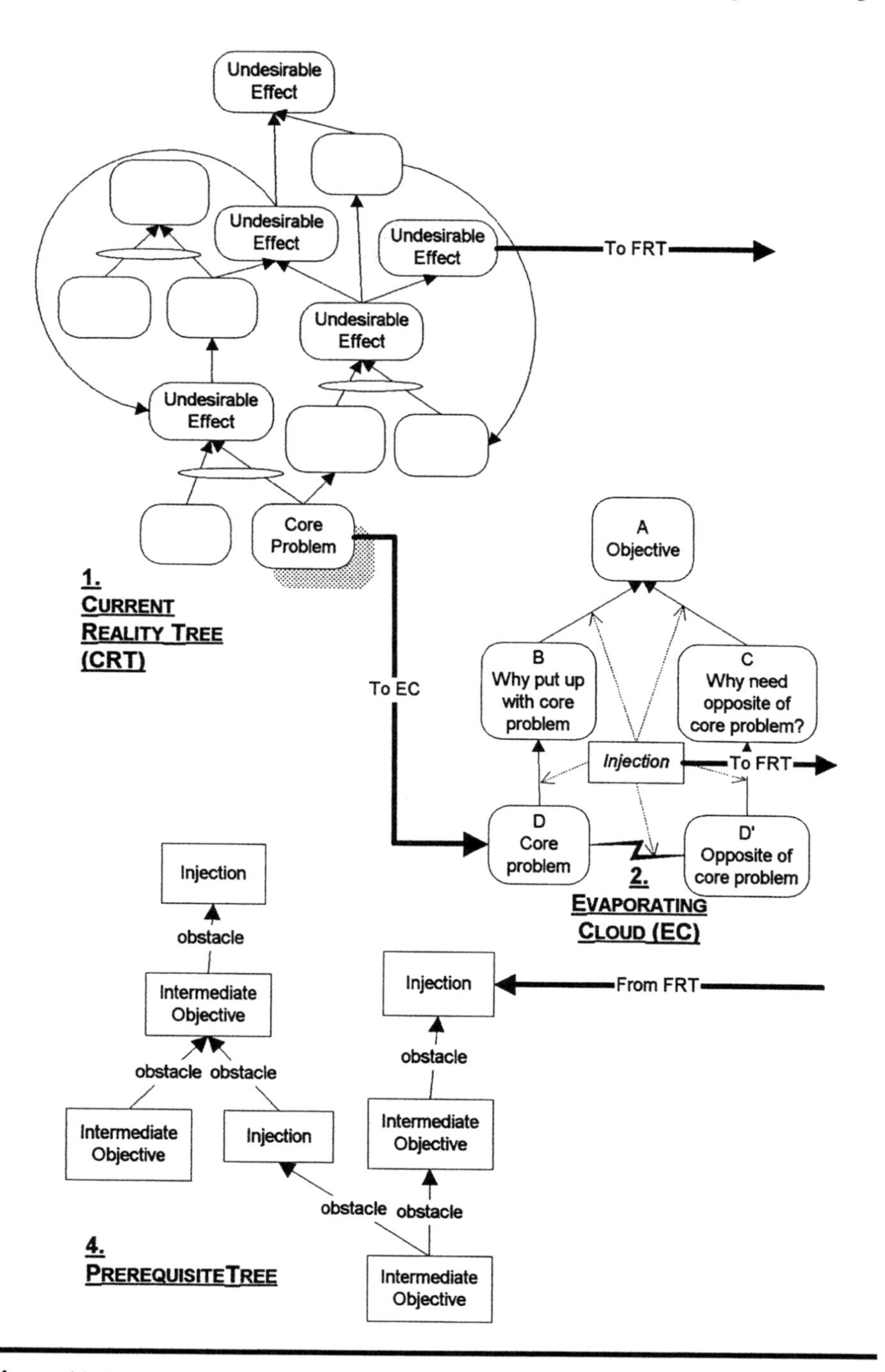

Undesirable Effect
Undesirable Effect
Undesirable Effect
To FRT
Undesirable Effect
Undesirable Effect
Core Problem
1.
CURRENT REALITY TREE (CRT)
To EC
A Objective
B Why put up with core problem
C Why need opposite of core problem?
Injection
To FRT
D Core problem
D' Opposite of core problem
2.
EVAPORATING CLOUD (EC)
Injection
obstacle
Intermediate Objective
obstacle
obstacle
Intermediate Objective
Injection
Injection
From FRT
obstacle
Intermediate Objective
obstacle
obstacle
Intermediate Objective
4.
PREREQUISITE TREE

Figure 11.5a

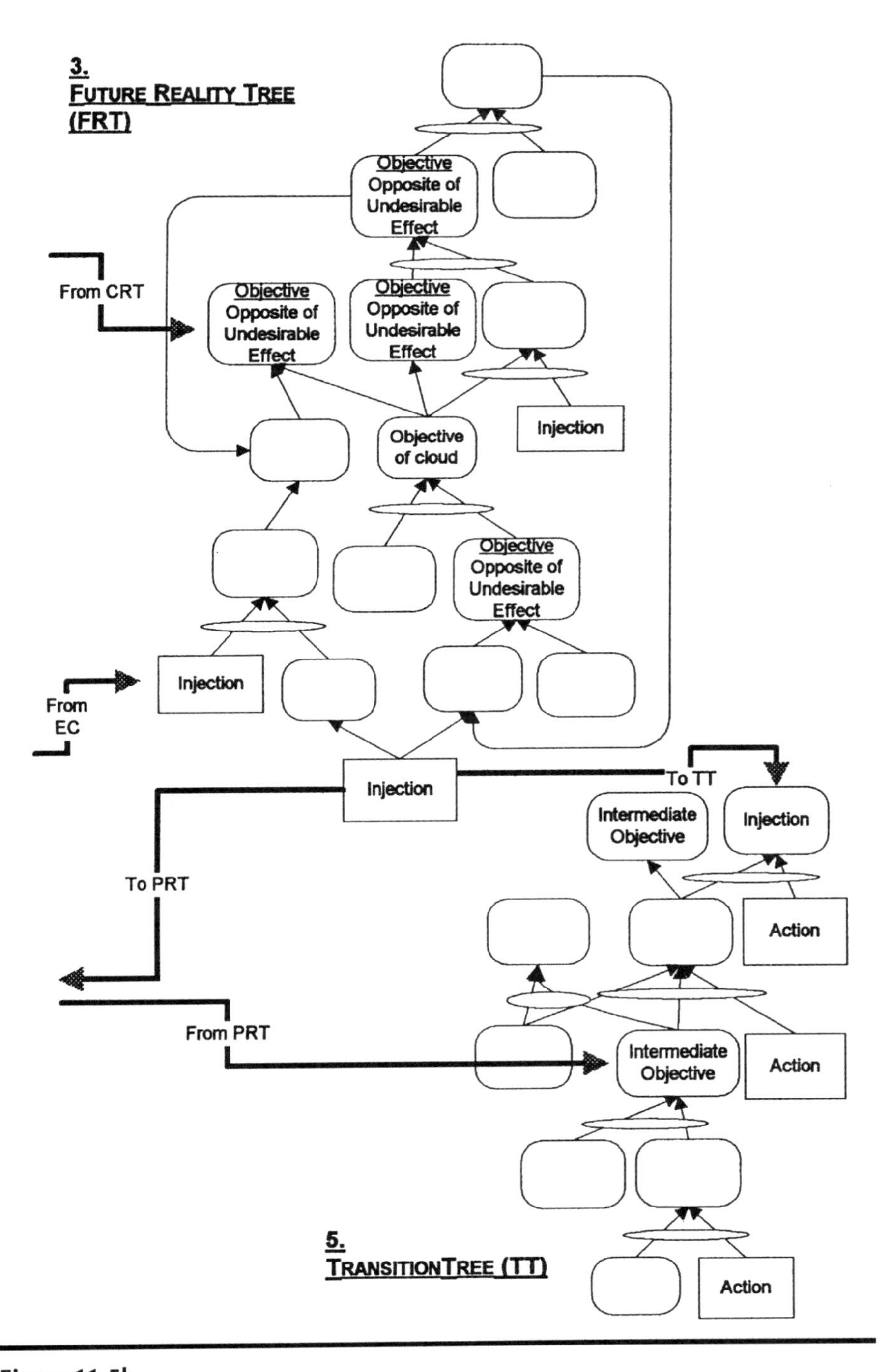
3.
FUTURE REALITY TREE
(FRT)
Objective
Opposite of
Undesirable
Effect
From CRT
Objective
Opposite of
Undesirable
Effect
Objective
Opposite of
Undesirable
Effect
Injection
Objective
of cloud
Objective
Opposite of
Undesirable
Effect
From
EC
Injection
Injection
To TT
Intermediate
Objective
Injection
To PRT
Action
From PRT
Intermediate
Objective
Action
5.
TRANSITIONTREE (TT)
Action

Figure 11.5b

Chapter 12

Communication CRT

There are any number of reasons that will lead you to want or need to communicate the content of a current reality tree that identifies a core problem. It may be that you want to enlist the help of a colleague to develop a solution to the core problem. Perhaps you need to get someone else's participation to implement a solution, and in order to get that participation, he must first understand the nature of the problem. What if the party to whom you need to communicate the content of the current reality tree is someone who is, or thinks he is, directly responsible for the environment described in the current reality tree? How do you go about communicating the issues to him without putting him on the defensive? If you have any experience in trying to explain problems, or in getting buy-in to solutions, you know that it is very difficult to be heard if the other party is feeling attacked. The communication current reality tree (CCRT) combines the evaporating cloud with the current reality tree for the purpose of communicating the current reality in a way that avoids the defensive response.

1. Convert the evaporating cloud into a sufficient cause diagram.
2. Connect this sufficient cause diagram to the current reality tree.
3. Communicate the new tree from the entry point that is the [A] entity from the evaporating cloud.

When you begin to create the communication current reality tree, you already have a completed current reality tree and evaporating cloud (Figure 12.1).

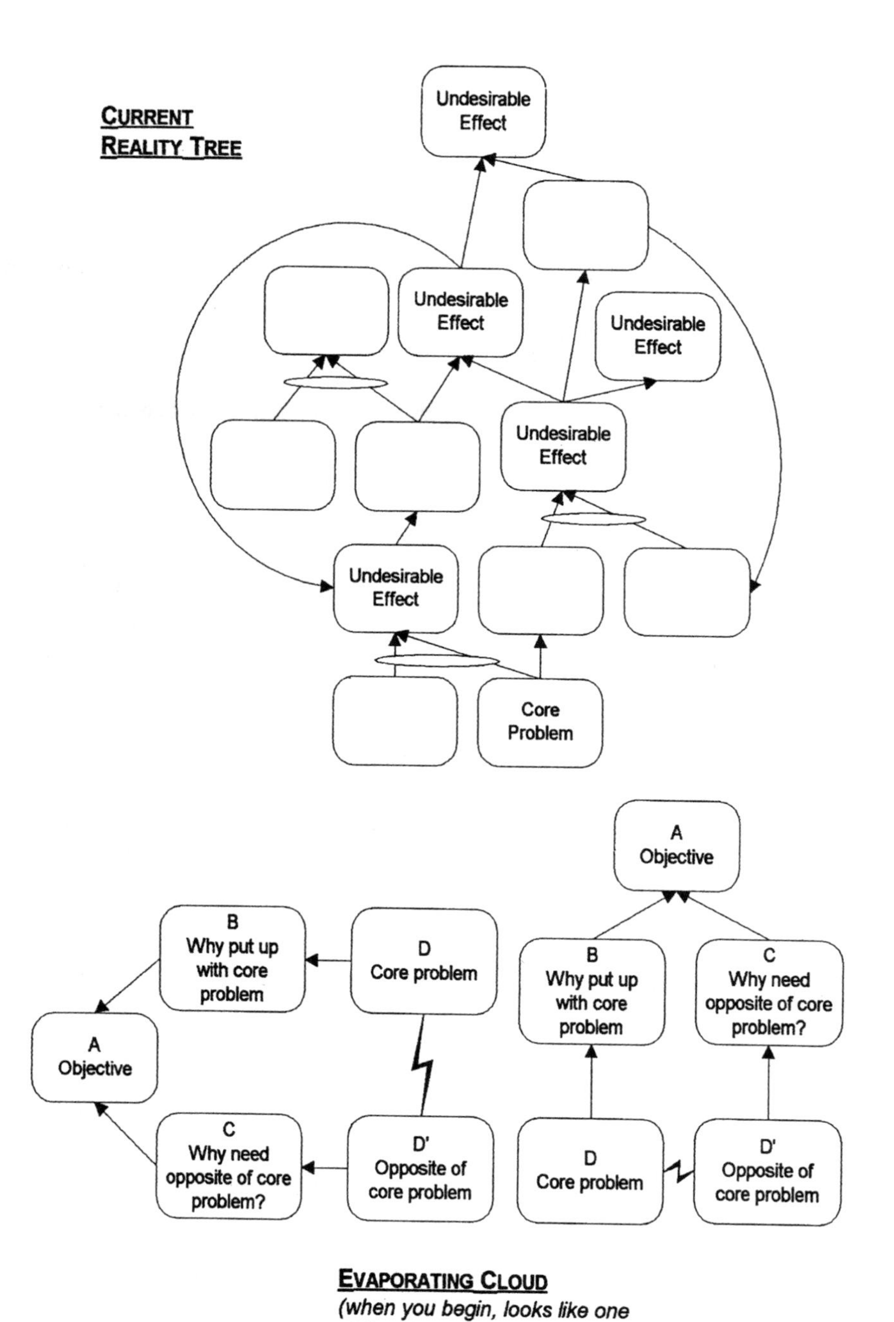
CURRENT REALITY TREE
Undesirable Effect
Undesirable Effect
Undesirable Effect
Undesirable Effect
Undesirable Effect
Core Problem
A Objective
B Why put up with core problem
D Core problem
C Why need opposite of core problem?
D' Opposite of core problem
A Objective
B Why put up with core problem
C Why need opposite of core problem?
D Core problem
D' Opposite of core problem
EVAPORATING CLOUD
(when you begin, looks like one of the above)

Figure 12.1

1. **Convert the evaporating cloud into a sufficient cause diagram.** Using the assumptions surfaced during the evaporating cloud process, turn the evaporating cloud into a sufficient cause diagram. The [A] objective will be an entry point to the new tree.
   - Because you will eventually be connecting this diagram to the current reality tree, the first step is to arrange the cloud's entities so once you turn it into a sufficient cause structure, the arrows will be oriented in the same direction as the current reality tree.
   - Using the assumptions of the cloud as your reference, turn the evaporating cloud into a sufficient cause diagram. A generic sample is illustrated in Figure 12.2.
2. **Connect this sufficient cause diagram to the current reality tree.** The [D] entity on the evaporating cloud is the same as the core problem, so simply connect it at the same place the core problem is shown on the tree. You will also be able to identify additional entities on the current reality tree that are caused by the conflict and the pressure or frustration involved in dealing with it. If the current reality tree does not already show it, this is an opportunity to expand it to identify that the lack of or compromised amount of the "D′" entity results in another need of the system. Two generic examples of communication current reality trees are shown in Figures 12.3 and 12.4.
3. **Communicate the new tree from the entry point that is the [A] entity from the evaporating cloud.** Before I go into detail on this step, I feel it is important to point out that there is much written on the subject of teamwork, communication, buy-in, collaboration, the psychology of change, etc. That is not what this book is about, although the categories of legitimate reservation are certainly helpful tools in that arena. This book is about a set of tools that, when combined with good intentions and basic communication skills, will help you improve whatever you set out to improve. The process described in this step is just a series of steps that have worked for some people. However, I caution you against putting these steps into action if your intent is manipulation. I caution you against putting these steps into action if you are not prepared to listen as well as share your analysis. The best advice I've ever read on the subject of communication was Steven Covey's: Seek first to understand before you seek to be understood*. Please keep Covey's advice in mind as you proceed.

* Covey, Stephen R., *The 7 Habits of Highly Effective People,* Simon and Schuster, 1989.

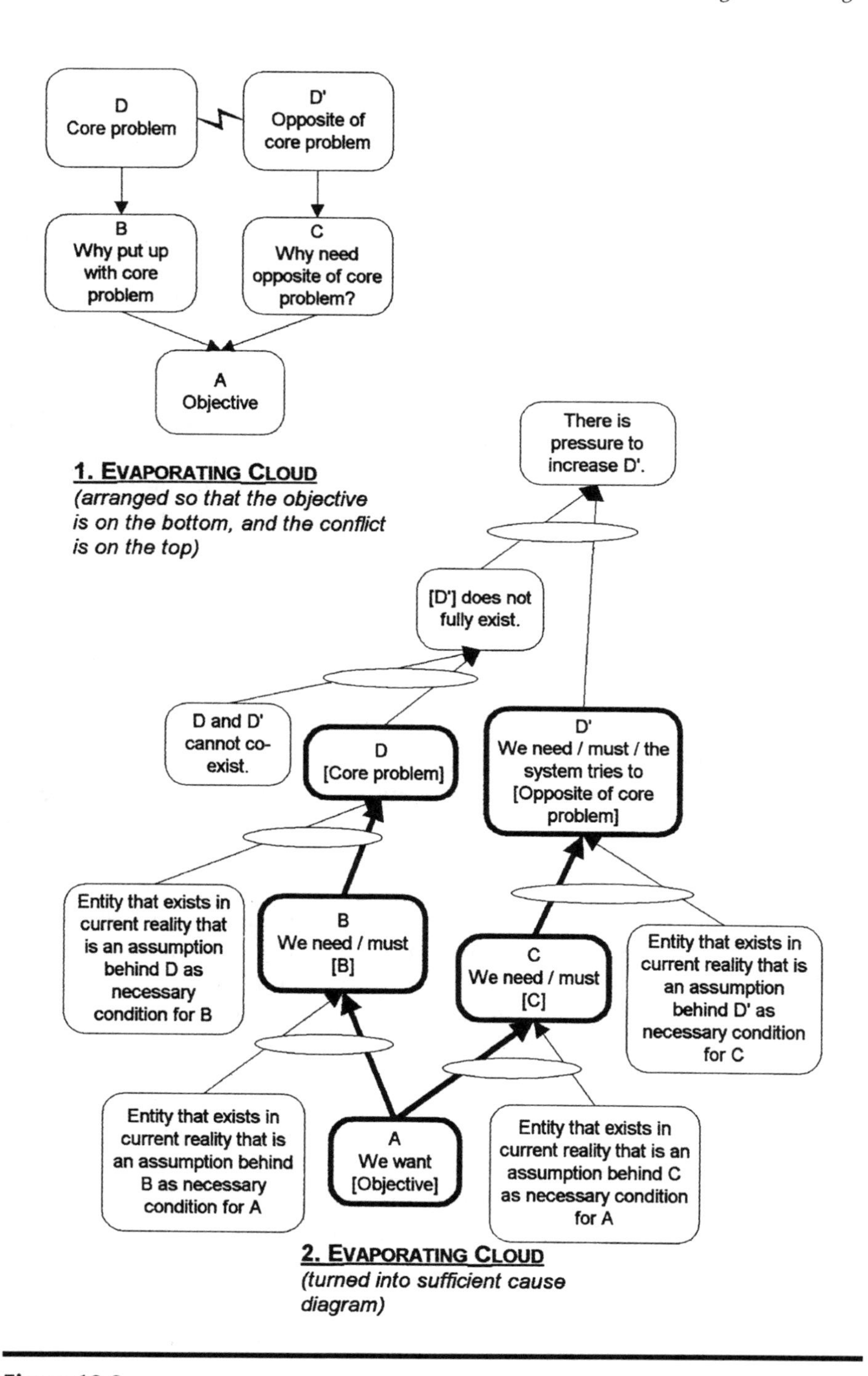
D
Core problem
D'
Opposite of core problem
B
Why put up with core problem
C
Why need opposite of core problem?
A
Objective
1. EVAPORATING CLOUD
(arranged so that the objective is on the bottom, and the conflict is on the top)
There is pressure to increase D'.
[D'] does not fully exist.
D and D' cannot co-exist.
D
[Core problem]
D'
We need / must / the system tries to [Opposite of core problem]
Entity that exists in current reality that is an assumption behind D as necessary condition for B
B
We need / must [B]
C
We need / must [C]
Entity that exists in current reality that is an assumption behind D' as necessary condition for C
Entity that exists in current reality that is an assumption behind B as necessary condition for A
A
We want [Objective]
Entity that exists in current reality that is an assumption behind C as necessary condition for A
2. EVAPORATING CLOUD
(turned into sufficient cause diagram)

Figure 12.2

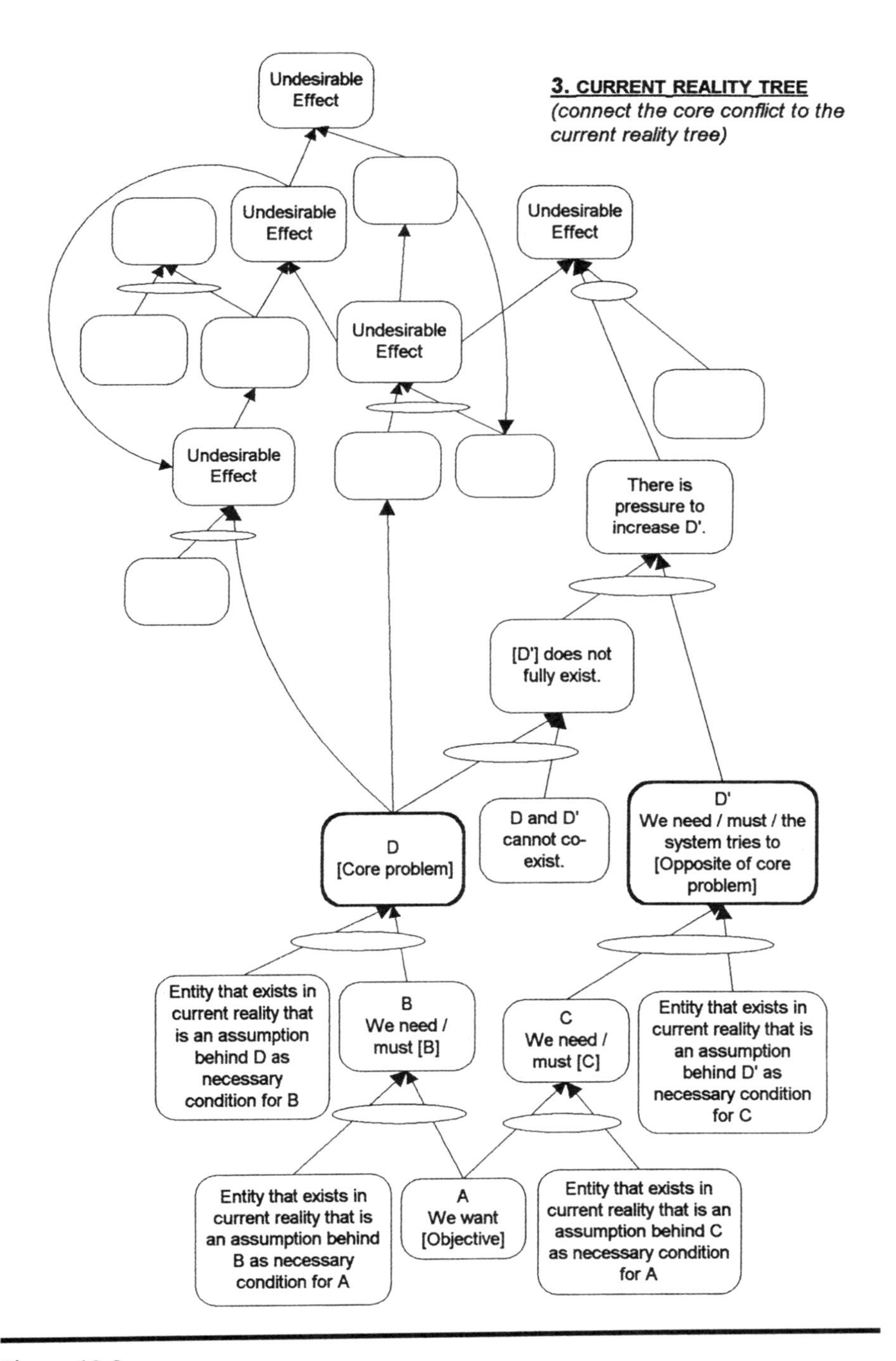

3. CURRENT REALITY TREE
(connect the core conflict to the current reality tree)
Undesirable Effect
Undesirable Effect
Undesirable Effect
Undesirable Effect
Undesirable Effect
There is pressure to increase D'.
[D'] does not fully exist.
D [Core problem]
D and D' cannot co-exist.
D' We need / must / the system tries to [Opposite of core problem]
Entity that exists in current reality that is an assumption behind D as necessary condition for B
B We need / must [B]
C We need / must [C]
Entity that exists in current reality that is an assumption behind D' as necessary condition for C
Entity that exists in current reality that is an assumption behind B as necessary condition for A
A We want [Objective]
Entity that exists in current reality that is an assumption behind C as necessary condition for A

Figure 12.3

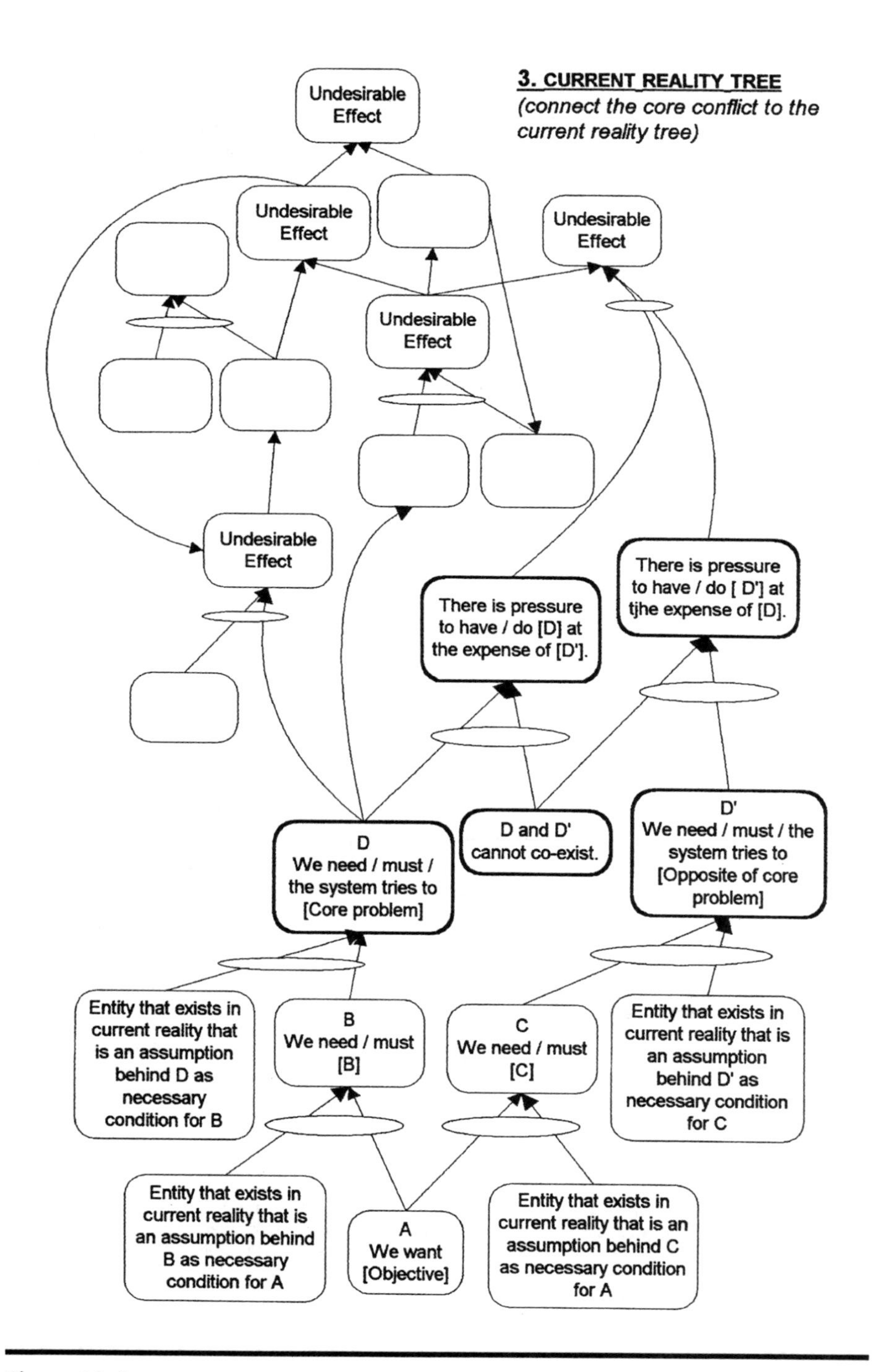
3. CURRENT REALITY TREE
(connect the core conflict to the current reality tree)
Undesirable Effect
Undesirable Effect
Undesirable Effect
Undesirable Effect
Undesirable Effect
There is pressure to have / do [D] at the expense of [D'].
There is pressure to have / do [D'] at tjhe expense of [D].
D We need / must / the system tries to [Core problem]
D and D' cannot co-exist.
D' We need / must / the system tries to [Opposite of core problem]
Entity that exists in current reality that is an assumption behind D as necessary condition for B
B We need / must [B]
C We need / must [C]
Entity that exists in current reality that is an assumption behind D' as necessary condition for C
Entity that exists in current reality that is an assumption behind B as necessary condition for A
A We want [Objective]
Entity that exists in current reality that is an assumption behind C as necessary condition for A

Figure 12.4

- Begin your conversation or presentation by establishing the common objective that is the stated objective of the cloud. This tells your colleague/s that you realize that you are all after the same thing. Make sure there is agreement that this is, in fact, a common objective. If he disagrees with you here, don't proceed. If it's the wording, change it. If he just doesn't agree that this is a common objective, it's better to stop, go back and reassess your analysis, and come back another time. Let's assume, however, that your colleague does agree that the objective is shared.
- Establish the [B] and [C] entities, as inevitable results of the existence of [A] and the entities that you have identified on the tree that combine with it to cause [B] and [C].
- Establish the [D] and [D′] entities in the same manner.
- Discuss the fact that [D] and [D′] are at odds — they are unable to coexist in the system. Invite your colleague to identify some of the problems that result from this conflict. No doubt, he will identify at least some of the undesirable effects that are on the current reality tree. The two of you have also now established that the inherent conflict is a system issue, and no single person, especially your colleague, is to blame.
- It is now OK to proceed with sharing the rest of your current reality analysis. Remember to keep the categories of legitimate reservation handy in your mind, so that as your colleague raises questions or concerns over what you've got in the tree, you can clarify, expand, or make changes as necessary.

Closing Comments

My intent with this book was to provide you with a user-friendly set of guidelines to the thinking processes. Some may wonder why there are not many more examples of trees in the book. It is certainly not due to lack of material. I made the decision not to litter the pages of the book with multitudes of trees because I wanted to focus on process rather than the content of what others had done in the past. I tried to put myself in the position of someone going to the bookstore and picking up the book, deciding whether or not to purchase it. As I envisioned scanning the pages and seeing all of those boxes and arrows, the thoughts that came to mind first were — cluttered, boring. I was reminded of many of the thinking process classes that I teach. In general, the participants want to get on with the process, rather than read through a large example. They want just enough to help make the process itself clear. With these thoughts in mind, I quickly saw myself in the shoes of the potential buyer at the bookstore, and I was not buying the book. Well, without the book in the hands of the buyer, it wouldn't be able to accomplish its main goal of being a learning resource. So, I decided on using what's worked in the past, and to put just enough examples in the book to facilitate the process.

If you are interested in examining more trees and evaporating clouds, all is not lost. APICS (the nonprofit educational society for resource management) maintains an internet list serve. The list serve is an e-mail discussion group. Simply join the list by sending an e-mail message to admin@lists.apics.org. In the body of your e-mail simply say, join cmsig. You will start receiving e-mail messages from other members of the list, and you can join in the discussion whenever you please. One of the things you can ask for is copies of trees that members have created, on whatever subject interests you. My hope is to one day see a virtual library of trees on the Internet — a site where we will be able to deposit copies of trees that we create and select and download trees that others have created.

The thinking processes have helped me in all aspects of my life. I hope that you, too, will use them, play with them, continue to find new applications for them, and, most of all, continuously improve with them.

> Our problems are man-made, therefore they may be solved by man. And man can be as big as he wants. No problem of human destiny is beyond human beings.
>
> John F. Kennedy, 1963

> Mind is the great lever of all things; human thought is the process by which human ends are ultimately answered.
>
> Daniel Webster

Appendix

Some Hints to the Nine Dots Puzzle

One common necessary condition relationship that people perceive is *In order to solve the puzzle, I must keep my writing instrument within the square*. This was not stated in the instructions, but because of the shape we see when we look at the dots, we assume that we must stay within that shape. Can you solve it if you break the arrow?

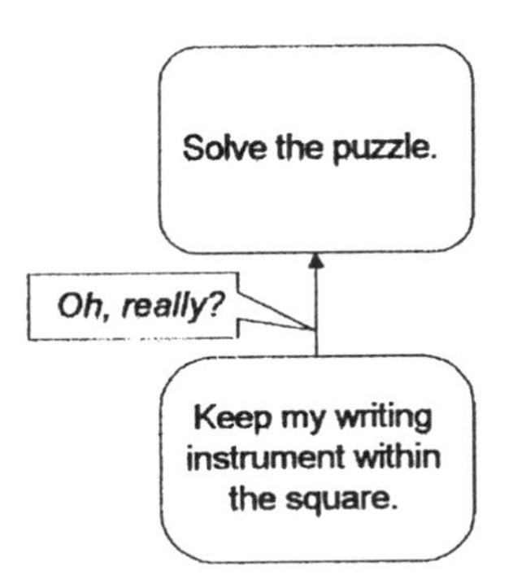

In order to solve the puzzle with three, two, or one line, the following necessary condition relationships should be challenged:

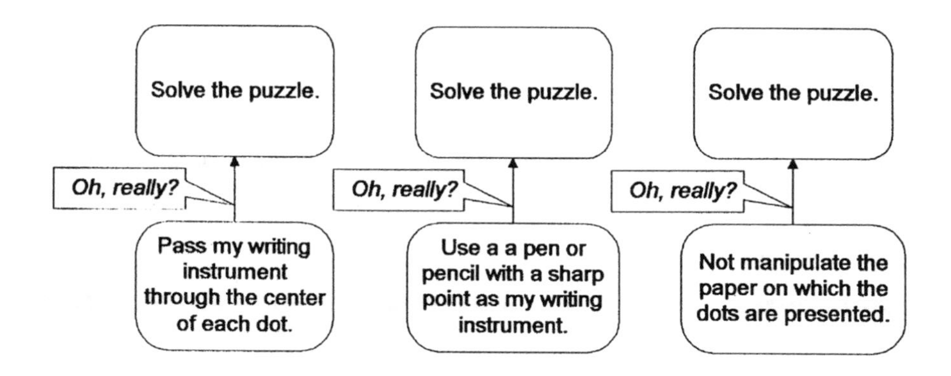

Some Answers to the Nine Dots Puzzle

Here are some answers to the nine dots puzzle. I'm sure that by identifying and challenging your assumptions, you will be able to create at least one more solution to the puzzle!

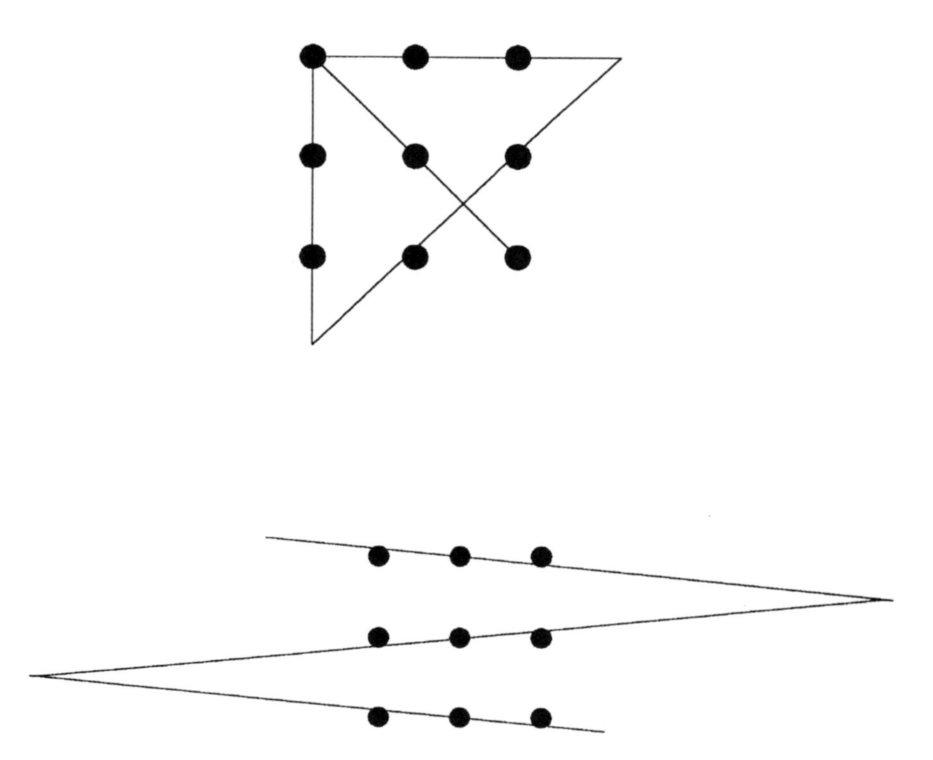

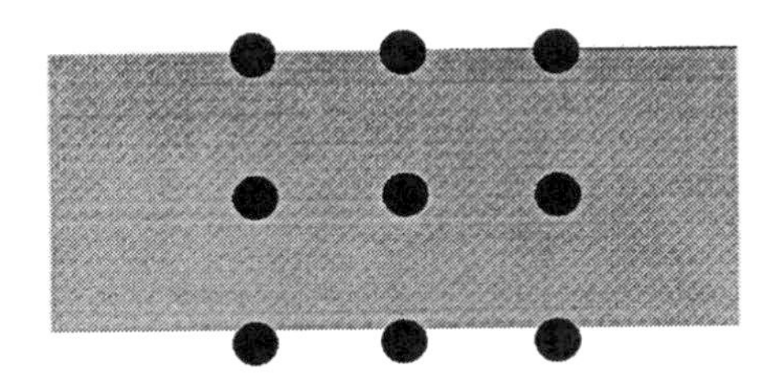

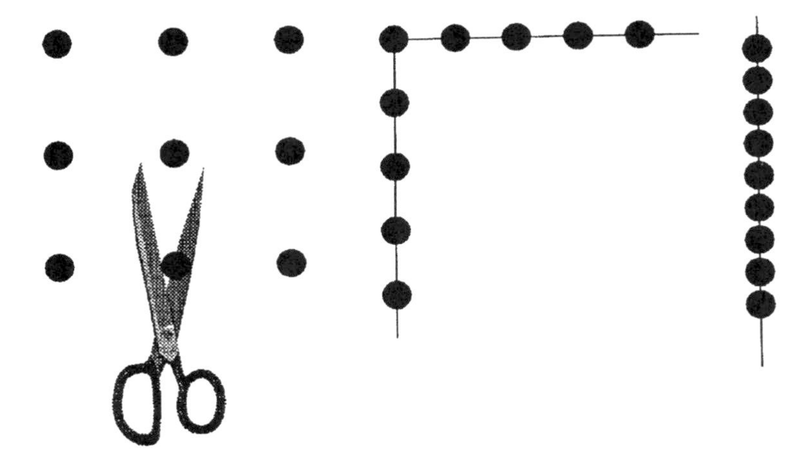

Index

D

E

F

G

H

I

J

L

M

N

O

P

Q

R

S

T

U

V

W

Y